Tourisr

ONE WEEK LOAN

New f .... itially focused on nature and
ecotou .... shift towards acknowledgement of
the soc .... ourism. The term 'sustainable tourism' has guided the
implic .... sustainability can be achieved, which may have led to the unfortunate belief
that sustainable tourism is impact-free. The term 'responsible tourism' is now gaining
popularity in part because it implies that all tourisms have impacts. This book discusses
the responsibility, or otherwise, of tourism activities in Latin America and the Caribbean. It
considers issues such as the reduction of poverty through tourism and the conflict between
increasing volumes of air travel spent in our continuing search for pleasure and the resulting
contribution to global warming.

Two major themes run through all the chapters: power and development. The authors
believe that tourism can be adequately assessed only through a consideration of how it fits
into the structure of power. It is also argued that tourism cannot be analysed without a
consideration of its impacts on and links with development. This relationship between
tourism, responsibility, power and development is explored in chapters covering both the
macro level of responsibility (international politics and tourism) and the micro level of
responsibility (local politics, poverty and tourism). The issues around the environmental
impacts of tourism, indigenous peoples and tourism, urban tourism and sexual exploitation
through tourism are also explored in detail. The authors look at methods of practising tourism
responsibly or irresponsibly at the personal, company, national and international levels. The
questions and dilemmas of 'placing' responsibility in the tourism industry are examined
throughout the book.

The book illustrates all these themes and issues as widely as possible with examples and
case studies from throughout the subcontinent, some documented nowhere else in the
literature on tourism, and, where appropriate carries the voices of the local people involved.
The book will be of importance to students and academics and to the work of practitioners
of development and tourism-related projects run by both governmental and non-governmental
aid and development agencies.

Martin Mowforth is a part-time lecturer in the School of Geography at the University of
Plymouth where his work focuses on issues of environment, development, sustainability,
natural disasters and tourism. He has been and still is an occasional development worker in
the region of Central America.

Clive Charlton is a principal lecturer in the School of Geography at the University of
Plymouth with a long-standing teaching and research interest in Latin America (especially
Mexico). His work focuses on issues of environment, transport, tourism and development.

Ian Munt is an independent urban development consultant and has worked on projects with
UN agencies, bilateral donors and non-governmental organisations in Central America,
Africa, Asia and the Pacific, and Europe.

Tourism and Responsibility

Perspectives from Latin America
and the Caribbean

**Martin Mowforth, Clive Charlton
and Ian Munt**

Routledge
Taylor & Francis Group

LONDON AND NEW YORK

First published 2008
by Routledge
2 Park Square, Milton Park, Abingdon, Oxon OX14 4RN

Simultaneously published in the USA and Canada
by Routledge
270 Madison Ave, New York, NY 10016

Routledge is an imprint of the Taylor & Francis Group, an informa business

© 2008 Martin Mowforth, Clive Charlton and Ian Munt

Typeset in Times New Roman
by Keystroke, 28 High Street, Tettenhall, Wolverhampton

Printed and bound in Great Britain
by Antony Rowe Ltd, Chippenham, Wiltshire

British Library Cataloguing in Publication Data
A catalogue record for this book is available from the British Library

Library of Congress Cataloging in Publication Data
Tourism and responsibility : perspectives from Latin America and the Caribbean / by Martin
Mowforth, Clive Charlton, and Ian Munt.
p. cm.
Includes bibliographical references and index.
1. Tourism–Economic aspects–Latin America. 2. Tourism–Economic aspects–
Caribbean Area. 3. Tourism–Government policy–Latin America.
4. Tourism–Government policy–Caribbean Area. 5. Tourism–Environmental
aspects–Latin America. 6. Tourism–Environmental aspects–Caribbean Area.
I. Charlton, Clive. II. Munt, Ian. III. Title.
G155.L3M69 2007
338.4'7918—dc22
2007013327

ISBN10: 0–415–42364–3 (hbk)
ISBN10: 0–415–42366–X (pbk)
ISBN10: 0–203–93440–7 (ebk)

ISBN13: 978–0–415–42364–9 (hbk)
ISBN13: 978–0–415–42366–3 (pbk)
ISBN13: 978–0–203–93440–1 (ebk)

Contents

List of tables, boxes and figures   vii
Foreword   ix
Acknowledgements   xi

1   Introduction   1

2   Global politics, power and play: the macro level of
    responsibility   10

3   Local politics, poverty and tourism: the micro level
    of responsibility   53

4   Tourism and the environment: eco by name, eco by nature?   101

5   Indigenous peoples and tourism in Latin America and
    the Caribbean   137

6   Urban tourism: the heart of darkness?   171

7   Sexual exploitation through tourism   195

8   Power and responsibility in tourism: know your place   224

    Appendix   Websites related to travel and tourism   231

Bibliography   234
Index   236

4.2   Costa Rica's Certification for Sustainable Tourism (CST)   114
4.3   The Monarch butterfly in Mexico   124
4.4   Carrying capacity and conservation of the Galápagos Islands   126
4.5   Colombian coffee tours   131
4.6   Mountain biking in Mexico   132
5.1   Selected articles of the Kuna Statute on tourism in Kuna Yala 155
6.1   Dwek on favela tours   188
6.2   Tourist guesthouses in Rio's shantytowns   190
7.1   Central American children's route into sex tourism   204
7.2   Tourism and sex in Ciudad Juárez   206
7.3   Evidence of sexual health problems   211
7.4   WTO/OMT requests to governments regarding sex tourism   216
7.5   WTO/OMT appeals to the travel trade regarding sex tourism   217
7.6   Extracts from TUI Nordic's hotel contract   219

Figures

1.1   Tourist arrivals to and migrants from LAC countries   6
2.1   History repeats itself . . .   15
2.2   All-inclusives and the local environment   25
2.3   Globalisation and the tourism industry   30
3.1   Poverty and tourism   58
3.2   Hotel Moka, Las Terrazas, Cuba   81
3.3   Trickle-down theory explained   84
3.4   Marina Puesta Del Sol, Nicaragua   86
3.5   Conflicts over water resources in Guanacaste, Costa Rica   91
4.1   Certificate of greenwash   116
4.2   Carbon budget calculations for selected short, medium and
      long haul flights   119
5.1   Indian graffiti in Puyo, Ecuador   140
6.1   Tourist groups guided around the restoration of Old Havana   183

Foreword

Q: 'What's the difference between ecotourism and mainstream tourism?' asked the tourist.

A: 'About 20 per cent' answered the tour operator – referring of course to his profits, but showing no recognition of the need for responsibility towards the environment, society, culture or community.

So perhaps when Norbert Suchanek suggested in the German-language *Sustainable Travel* magazine that 'Ecotourism is dead . . .', he captured the mood of all those 1990s ecotourism enthusiasts and promoters who came to realise that their 'eco-' prefix needed either to be recaptured or replaced. And so, towards the end of last century's last decade and more recently, the new forms of the tourism industry reinvented themselves with new labels: community-based tourism, pro-poor tourism and responsible tourism being foremost among them.

For the purposes of this book, we have chosen the term 'responsibility' and its association with the tourism industry as a central theme owing to its generic nature and the fact that it does not restrict our coverage of the many not-so-new forms of tourism. In fact, this is an issues-based book whose underlying theme of responsibility – individual, governmental and international – lies at the heart of philosophy, as Socrates and many others have made clear. But we do not delve deeply into philosophy – we are neither qualified nor able to do so. Instead, our examination of human responsibility is made through the prism of the tourism industry, and in reality our analysis is more of a series of illustrations rather than a philosophical treatise.

We would like to have chosen many other issues associated with the tourism industry, but for a variety of reasons and restrictions our coverage has to be limited. Through the few issues that we have examined, however, it has become clear to us that the underlying theme of responsibility is reflected through all forms of the industry and in myriad ways.

As an issues-based book, it is not designed to be replete with examples of good practice of responsible tourism that either gives the impression that tourism is a wholly beneficial activity or serves as a manual of how to practise tourism 'correctly'. Rather, it is intended as a discussion of responsibility in tourism, which of course implies that coverage has to extend to the other face of the issue, namely *ir*responsibility.

the work of Anita Pleumarom. Both organisations are constant sources of inspiration as well as information. We also express our thanks to a number of other organisations which include Survival International, AVERT, the World Tourism Organisation (Madrid), the Institute for Central American Studies (San José, Costa Rica), North American Congress on Latin America (NACLA, New York), Tearfund (UK), the International Porter Protection Group (IPPG), and Progressio (formerly the Catholic Institute for International Relations). Additionally it should be noted that the idea for the book first came from the Latin America Bureau in London.

The work of the Cartography Unit in the School of Geography at the University of Plymouth deserves special mention for its professionalism and high quality. We are particularly grateful to the Unit's staff, Brian Rogers, Tim Absalom and Jamie Quinn – Jamie was responsible for producing the majority of the Figures and Tables in the book. At Routledge, Andrew Mould and Jennifer Page have been very encouraging and we are grateful to them for their patient assistance and in particular for accepting and gaining approval for the book.

June, Ruth and Diana deserve our utmost gratitude for their forbearance, encouragement and patience.

Every effort has been made to contact copyright holders for their permission to reprint material in this book. The publishers would be grateful to hear from any copyright holder who is not acknowledged and will undertake to rectify any errors or omissions in future issues of this book.

1 Introduction

This book is a mixture. We try to pull together various themes relating to tourism and development in the continental context of Latin America and the Caribbean – from this point on referred to as the LAC countries. It is intended partly for those who have already visited or who wish to visit these countries – but it is not a guide book. It is an issues-based book, examining particularly those themes relating to the responsibility and appropriateness of travelling to the LAC countries and the role of tourism in the broader field of development. In this sense, it should also be of interest to a wide range of academics – those in development studies, anthropology, geography, sociology, economics, politics and the recently emerged study of tourism itself. But it is not an academic text. Additionally, it should be of interest to members of the governmental, non-governmental and supranational agencies which proactively plan and promote tourism developments in the LAC countries and elsewhere – but it is not a manual on how to practise or develop tourism.

In fact, we would not wish to promote the idea that there is a 'correct' way to tour, to service tourists, or to develop tourism. Yes, tourism can be practised in a relatively responsible, sustainable and ethical way. But the important word here is 'relatively', for the notions of responsibility, sustainability and ethics are relative to the values and perspectives of all those who participate in the activity of tourism. And these values and perspectives differ according to the respective roles played by participants. Members of indigenous groups, for example, are unlikely to perceive the behaviour of tourists in the same way as hoteliers or service providers in non-indigenous communities. Different groups of tourists perceive the responsibility of their tourist activities in differing ways. Some will claim that nature tourism is responsible even though some of the areas protected specifically for this activity may exclude local people. Others claim that all-inclusive resorts are responsible because they have minimum social and cultural impact on the local populations. The 'pure' nature tourist may scoff at the meanness of the low-budget backpacker, who in turn may scoff at the rich conventional tourist whose money is perhaps more likely to generate employment in the host country. Which of these types, and other types, is the most responsible? We do not offer an answer to this question. Rather, we simply wish to highlight these different perspectives and the fact that these notions are hotly contested.

Rather than declaring the actions and practices of different groups to be responsible, sustainable or ethical, it would probably be wiser to see these notions on a continuum: some activities may be more responsible than others, but nowhere and at no time are the actions of tourists, tourism developers or service providers without impacts. And all such impacts can be both negative and positive, these counterbalancing each other in varying degrees. To claim that a given activity is 100 per cent economically sustainable does not mean that sustainability has been achieved, for the notion of sustainability has many facets: not just economic, but also social, ecological and cultural. These different facets of sustainability are often traded off against each other. We believe, however, that these notions should be seen in their entirety, all elements integrated with each other in complex ways.

Additionally, it is important to state that the book is not just a collection of best case scenarios or good examples. Many of the examples here included can be seen as good in our judgement. But our judgements are made from the standpoint of white, middle-class professionals from a wealthy country. As such, we are members of the proportionately small (but growing) group of the planet's population who are able to travel relatively freely to most parts of the globe and on whose custom the growing phenomenon of international tourism is pretty much dependent. It is highly unlikely that our judgements and perceptions are shared by many other individuals and groups who experience tourism from different angles. And what is seen as benign by some may well be seen as malign by others. So the book includes a range of examples and discussion of the other side of these notions, namely irresponsibility, unsustainability and unethical behaviour or development.

It is important at the outset that we also give the reader an idea of our understanding of the notion of development, for our analyses of many examples, problems and issues relating to tourism are informed by this understanding. The word 'development' is imbued with ideas of progress and it is widely, almost universally, understood as a good thing. It is also imbued with a sense of inevitability. It has been this way since President Truman declared it so in 1949 and designated much of the world as 'under-developed' and in need of development. As Esteva (1992)[1] and Escobar (1995)[2] have it, on that day, 20 January 1949, two billion poor people were discovered and became under-developed. But it seems to us that the question of what form this development should take is rarely asked. Instead, it is widely presumed that it should be in our own image, that image being one of western capitalism. This may or may not be a good and wise form of development – the only game in town as Francis Fukuyama's 'The end of history?'[3] would have us believe – but it needs to be understood as only one model of development, albeit complex, rather than the only model of development. This implies of course that we should question the general notion of development and not accept it as automatically meaning something that is benign, nor as being inevitable. The current trend in Latin American democracies to register a popular distaste for such neoliberal policy measures as the privatisation of public utilities and cutbacks in public sector funding may not yet represent a fundamental questioning of the notion of development, but it does at least show signs that history may not yet have ended after all.

This idea of questioning the general notion of development does not necessarily mean that we view these issues from the standpoint of the maturing anti-globalisation movement. But it does mean that we accept that such issues are intrinsically political. As Cheryl Shanks puts it, however, in the field of tourism studies:

> Tourism pretends to be apolitical, but it encapsulates problems of power and worth on a grand and global scale; it pretends to be passive, yet it is produced by an encounter between host and guest in which anything is possible. When tourists encounter local people, they bring with them the weight of their expectations, their leisure and their power. Locals see this, and respond: they react against it, make a counter offer, or adapt to expectations. This seemingly trivial exchange can have profound economic, environmental, cultural and political effects, not only on individuals but on the global political economy.[4]

Acceptance of the political nature of development means that any analysis of development, and of tourism development, must be informed by an examination of the structure of power and privilege relating to decisions made about that development. And this can be taken as an implicit acknowledgement that these decisions will be made according to the values of those making them.

Put more simply, we see development, and tourism development and developments, as political and value-laden. We therefore try to examine the issues and examples in this book in this light. Who makes the decisions will determine, at least in part, who are the beneficiaries.

Given, then, that general development and tourism development are political in nature, what is the relationship between the two – tourism and development? As Caroline Ashley of the Overseas Development Institute suggests in referring to the effects of the Indian Ocean tsunami, 'the contribution of tourism to development is never so widely recognised as when tourism collapses'.[5] So, does tourism bring development? Or does development bring tourism? These questions underlie our examination of many of the examples and issues discussed in this book, and we find at times that we cannot avoid the relevance and importance of the question of whether tourism can ever be sustainable and responsible where the prevailing model of development is unsustainable and irresponsible. Efforts to develop tourism sustainably and responsibly face an uphill struggle in such an environment.

One further general issue associated with the notion of development that requires a little more discussion at the outset is its relationship with poverty and inequality. As awareness of globalisation and economic integration has grown, so has the idea that the beneficiaries of development will be or should be the poor. The United Nations' Millennium Development Goals make it explicit that, essentially, all development roads lead to the need to eliminate poverty. And it has been a recent feature of the field of tourism that relevant government agencies (such as the UK's Department For International Development, DFID) are now promoting tourism as a means of alleviating poverty and reducing inequality. We therefore also attempt to examine the issues and examples discussed in this book in this context, and where

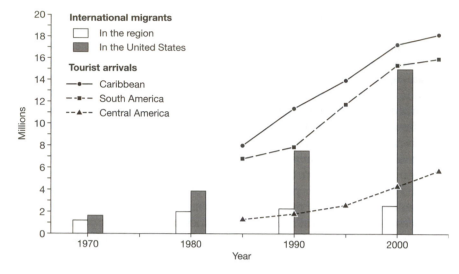

Figure 1.1 Tourist arrivals to and migrants from LAC countries

Sources:
Tourist arrivals: WTO 'Tourism Market trends', 2005 Edition (Data is collected by WTO November 2005).
Migration data: Economic Commission on Latin America and the Caribbean (ECLAC) notes, May 2002 (No. 22).

from which emigrants leave and those areas which receive tourists; again, however, equally there are also many and an increasing number of areas which serve as both generators of emigrants and receivers of tourists.

Lest anyone should misunderstand, let us also make it clear that there are increasing levels of wealth in the LAC countries at the same time as there are increasing levels of poverty. The two do not contradict each other. They occur together. Inequality is not manifest only on a global scale; it occurs within and between countries and in relation to a variety of characteristics, particularly sex, ethnicity and class. As Dodds comments: 'Put simply, there are parts of the Third World in the First World and *vice versa.*'[10] This has implications for tourism. There is a growing middle class in the LAC countries and they experience a growing global middle-class culture which stimulates them to aspire to the same kinds of holidays and tourism as the growing middle class of the rich world.[11] As we jet around the world on our tourist circuits, we might find it difficult to escape the conclusion that in the LAC countries tourism is an industry run by foreigners for the benefit of foreigners. But a more informed knowledge of the trends and patterns shows that the domestic middle classes of national populations are just as keen to take part in tourism as are the foreigners.[12]

We suspect that both trends, of growth in international tourism and international migration, reflect the tendency of globalisation to divide the world into two sectors, the haves and the have-nots. But this is too simplistic. The processes through which these trends have occurred are complex and multi-faceted and an understanding of them demands an inter-disciplinary and nuanced approach. In a short, discursive book such as this, it is impossible to deal with the complexity of these issues in sufficient depth, but we need to confirm their existence and we try to indicate some of their dimensions. In Chapter 2 we briefly cover the historical development of these trends and the global political forces (represented by supranational organisations such as the World Bank, the International Monetary Fund and the World Tourism Organisation) behind these trends and the national effects of these forces on tourism in the LAC countries. In Chapter 3 we extend the analysis of these global effects to the local level and examine the relationships between tourism and poverty and inequality in greater depth. In particular, we examine the meaning and extent of local participation in tourism developments. In Chapter 4 we ask questions about the links between tourism and the environment in the LAC countries. This involves issues and labels of sustainability and includes a brief encounter with the most recent associated techniques of ecological footprinting and carbon budget calculations, both directed specifically to the tourism industry. One potentially critical point of contact and conflict between tourists, the tourism industry and local groups which brings issues of responsibility into sharp focus concerns visits to tribal groups. Chapter 5 examines these contacts and conflicts and covers illustrative case studies rather more extensively and intensively than other chapters, in which we have tried to limit the depth of case studies in order to include a greater number of them. Chapter 6 deals with urban tourism. This is not just the conventional visiting of monuments, museums, palaces, parks, theatres and major sporting and cultural events but also the recently emerged practice of guided tours around the *favelas* and shanties characterised by poverty and crime. Such tours are still of minor significance in the tourism industry as a whole, but we ask where and how does responsibility fit into this type of tourist activity? Chapter 7 looks into yet another recently emerged issue of responsibility, or irresponsibility, in tourism, namely tourism for sexual gratification. This was previously associated with south-east Asia rather than the LAC countries, but some of the latter are rapidly gaining a reputation for sex tourism, widely seen as the height of irresponsibility in tourism. We delve into the issues surrounding this feature of international tourism as it affects the LAC countries.

Finally, in Chapter 8, we draw together a few lessons from the earlier chapters, not in a way that produces a list of DOs and DON'Ts, but rather as a means of showing that all these issues are related to each other. We raise the questions of why we travel to the LAC countries, who benefits, who loses and how is development, if we can define it, affected? We do not tell and do not want to tell people to travel or not to travel to specific places or for specific reasons. But we do try to raise some questions which we think all travellers to the LAC countries should ask themselves before they go, whilst they are there and after they return home.

Notes

1 Esteva, G. (1992) 'Development', in W. Sachs (ed.) *The Development Dictionary: A Guide to Knowledge as Power*, London: Zed Books.
2 Escobar, A. (1995) *Encountering Development*, Princeton: Princeton University Press.
3 Fukuyama, F. (1989) 'The End of History?', *The National Interest*, Summer: 1–18.
4 Cheryl Shanks (2002) 'Tourism in the Americas: Nine Quandaries of Tourism', *Revista: Harvard Review of Latin America*, Winter.
5 Caroline Ashley (2005) 'The Indian Ocean Tsunami and Tourism', *ODI Opinions* 33, Overseas Development Institute (London), January.
6 The World Bank (annual) *World Development Report*, Oxford University Press.
7 Reuters, 1 June 2006, in Reuters AlertNet, www.alertnet.org/thenews/newsdesk/ accessed 23 July 2006.
8 Giles Tremlett (2006) 'Latin American Migrants Send Home £27bn', London: *Guardian*, 15 November 2006.
9 In this book we refer specifically to the LAC countries, but there are occasions when the alignment of the LAC countries with other parts of the world cannot be avoided. Where that is the case, we generally refer to *Third World* and *First World*. The history of development studies has thrown up a variety of terms that attempt to represent and categorise countries according to their wealth and social wellbeing. In particular, the terminology attached to countries lower down the 'human development index' (a widely adopted index ranking countries on a number of criteria) has been keenly disputed – should they be described as 'poorer', 'lower income', 'developing', 'under-developed', 'the South', 'Third World', or indeed 'non-viable economies' (De Rivero, 2001) or 'slow economies' (Toffler, quoted in Sachs, 1999)? All such terms possess their advocates and detractors, reflect political priorities and dispositions, and no one term will suit all audiences. De Rivero, O. (2001) *The Myth of Development: The Non-viable Economies of the 21st Century*, London: Zed Books. Wolfgang Sachs (1999) *Planet Dialectics: Explorations in Environment and Development*, London: Zed Books.
   The term 'Third World' helps to reflect on and convey the way in which we are using this term in relation to the notion of development. For example, the word 'developing' is avoided, because it implies that there is an end state to the process of development and that all countries will eventually reach a 'developed' state. By contrast, there are strong grounds for arguing that the process of development is one which actually causes under-development elsewhere and at the very least is a state of never-ending flux to the extent that all countries are 'developing'. The term 'Third World', in other words, helps to emphasise the ways in which power, resources and development are unequally and unevenly shared globally – if anything the very term 'Third World' requires an acknowledgement that, despite fifty or more years of 'development' activity, profound global inequalities both persist and are increasing. This is not to say, however, that the Third World is easily defined as a neat geographical entity coterminous with nation states.
   Although strict geographical divisions along national borders are increasingly meaningless, as a generalisation in talking of the First World we are referring to the power vested in the nation states and institutions of North America, Europe, Australasia, and Japan. The Third World refers to those nation states and institutions that make up Latin America, the Caribbean, Africa and parts of Asia, although it is necessary to acknowledge the increasing wealth of some countries in South Asia and the Pacific Rim.
10 Dodds, K. (2002) 'The Third World, Developing Countries, the South, Poor Countries', in V. Desai and R. Potter (eds) *The Companion to Development Studies*, London: Arnold, p. 6.
11 Whilst we acknowledge that there has been a retreat from the analysis of social class at

the same time as a rise in the postmodern analysis of sociological phenomena through individualism – so that many people have begun to think in terms of their individual status rather than their class status – we consider that the relatively recent recognition of class fractions, especially within what is referred to as the new middle classes, offers a means of analysis of great pertinence for tourism. This is witnessed particularly in new and responsible forms of tourism, which serve as life features through which groups of people distinguish themselves from other groups and thereby maintain a social distance from others. Thus, certain types of tourism develop into a fashion which certain groups follow until it becomes trendier to characterise oneself by a different type. In this 'game', tour operators become both followers and setters of fashion, choosing to promote both types of tourism and destinations to particular social groups – or class fractions.

Bourdieu's work on the notion of *habitus* – the appropriation of objects, such as holidays, by certain social groups in order to differentiate themselves from other social groups – is crucial to this issue and can be found in: Bourdieu, P. (1984) *Distinction: A Critique of the Judgement of Taste*, London: Routledge and Kegan Paul; Bourdieu, P. (1987) 'What Makes a Social Class? On the Theoretical and Practical Existence of Groups', *Berkeley Journal of Sociology* 22: 1–17.

In this book we use the very general term 'middle class', but we acknowledge the importance to tourism studies of class fractions and we refer the reader who wishes to pursue this line of analysis to the discussion of 'Class, Capital and Travel' in Chapter 5 of Mowforth and Munt (2003) *Tourism and Sustainability: Development and New Tourism in the Third World*, Second edition, London: Routledge.

12 There may be some difference between domestic tourists and foreign tourists in terms of their type and characteristics of tourist activity. For instance, it is possible that domestic tourists are more likely to be day trippers than long-stay visitors and certain destinations may be more associated with domestic visitors than with foreign visitors. Moreover, we acknowledge that the incidence of domestic tourism in the LAC countries is a growing phenomenon.

For further reading on this matter we refer the reader to Krishna B. Ghimire (ed.) (2001) *The Native Tourist: Mass Tourism within Developing Countries*, London: Earthscan.

2 Global politics, power and play: the macro level of responsibility

It is perhaps a mistake to make the first substantive chapter of the book one that deals with the historical development and political economy of Latin American and Caribbean countries – it can be a dry subject at times, one that involves many 'isms' and 'ations' – privatisation, regulation, globalisation, corporation, neoliberalism. As organisations such as Oxfam, Christian Aid and the World Development Movement are fully aware, it is not easy to engage and retain the interest and enthusiasm of the populace at large for issues that can seem distant, with interlinkages that are complex and difficult to understand. But the development of the structure of power and privilege within the LAC countries and the history of the exercise of power over the region from outside forces are crucial to an understanding of the influences on the development of tourism. The same factors are no less relevant to an analysis of the responsibility or otherwise with which tourism developments are made.

We take the stance that the development of tourism is not apolitical. This means that the issues relating to the ethics and values of practising tourism responsibly are not politically neutral. The conservation of nature for the benefit of tourists, the construction of a tourist enclave in a community which depends on other activities, the introduction of western fast food outlets into a culture which has traditionally treated food in a significantly different manner – these activities and many more are all political and represent changes in the balance and exercise of power.

Briefly, we examine the historical development of tourism in Latin America and the Caribbean. We ask whether the LAC countries are changing in nature from the provider of primary products ('banana republics') to becoming the playground of the wealthy who come mostly from the First World ('playground republics'). We examine the background notion of development and the meaning of under-development as it affects the majority of the LAC countries; and this leads us on to the Millennium Development Goals and their relevance to the field of tourism. Necessarily this involves an examination of the insertion of tourism into the emerging markets of these countries, the neoliberal economic strategies pursued by national governments, associated labour practices, and the influences of international financial institutions (such as the World Bank and International Monetary Fund) and other supranational organisations such as the World Tourism Organisation. Transnational corporations (TNCs) are potentially highly influential in tourism development strategies and their role and practices are also examined. Additionally, we look into the political strategies associated with the dominant economic model of neoliberalism (privatisation, deregulation, free trade zones, the lowering and/or elimination of tariff barriers) as they affect tourism developments. And finally, we consider a number of concepts and techniques (such as voluntary self-regulation and corporate social responsibility) used to mitigate the ill effects of these strategies as they apply to the development of tourism.

A potted political history of tourism development in the LAC countries

Chapter 1 made brief reference to the early history of tourism in the LAC countries, beginning with the arrival of the Spanish conquistadors, following hot on the heels of Christopher Columbus. Perhaps it is pushing the definition a little too far to call these early visitors 'tourists', but they were certainly the agents who opened up the continent to others who followed, initially from Europe. A little later came the European slave traders and pirates (often mistakenly referred to as valiant explorers) who brought with them slaves from Africa.

These early visitors 'supplanted a vitally productive, self-contained and self-sufficient pre-Hispanic system of agriculture, which had supported sophisticated Indian cultures before the arrival of the conquistadors'[1] with the plantation, although it is important not to be too romantic about the earlier civilisations whose social and environmental levels of sustainability are the subject of much argument among history, archeology, anthropology and Latin American scholars. The plantations were serviced by an economic system of slavery and debt labour. Eventually the Spanish crown abolished slavery, but as Weinberg explains: 'the debt labour which still exists today on Central American plantations, has maintained a captive, dirt cheap labour force throughout the centuries'.[2] The plantation system evolved into the *latifundio*, which Eduardo Galeano describes as

> one of the bottlenecks that choke economic development and condemn the masses to poverty and a marginal existence in Latin America today. The *latifundio* no longer depends on the importation of African slaves or on the *encomienda* of Indians; it merely needs to pay ridiculously low wages, or to obtain labour for nothing in return for the labourer's use of a minute piece of land.[3]

Thanks in part to Simon Bolivar in South America and nascent nationalist and rebellious sentiments in the Caribbean, most of the LAC countries were granted independence from their European colonial visitors during the nineteenth and twentieth centuries. But by that time the dominant European culture had left its mark on the indigenous populations. In a century and a half after the first arrival of the conquistadors the Indian populations of the Americas had been reduced from 70 million to 3.5 million,[4] the result of genocide arising from the colonial arrogance and ignorance of the European visitors and the impact of imported diseases against which indigenous populations had little resistance. The rape of indigenous women had given rise to the creation of new ethnic groups of mestizo[5] and mulatto[6] populations.

The balance of population, economy, culture and power had been changed by these early visitors.

Independence from colonial rule did not shake off the yoke of domination from beyond. As European influence ebbed, North American influence began to rise. The shade of the military uniform of the visiting invaders changed, but the extraction of wealth and resources continued. The new military invaders could not be mistaken for tourists, but along with them came their companies and their companies' employees to continue the extraction of wealth and resources from the subcontinent. The first few editions of the *South American Handbook*[7] (first published in 1924 by Royal Mail) were used by business travellers to Latin America, especially those involved in mining, railways and cattle. A variety of places were turned into playground enclaves for the rich and powerful, under the protection of the client dictators, sometimes installed and nearly always supported by the US government, and their controllers of public order, the police and/or military forces. Havana, for example, was one of the favourite haunts of mafia figures from the USA until the Cuban revolution in 1959 ended its reputation as the sin city of the Caribbean. Managua too was known as a den of vice, much frequented by the rich of the north, before its centre was torn down by the 1972 earthquake. The beaches of Rio de Janeiro also attracted the rich playboys and beautiful hedonist jetsetters, and resorts such as Mar del Plata in Argentina and Punta del Este in Uruguay grew to serve hedonistic local elites.

In the type of development which characterised the region it is difficult not to see the background which inspired the dependency theory of development which describes some parts of the globe as marginalised and dominated by other parts in order to fuel growth and development elsewhere. The theory argues that western capitalist countries have grown as a result of the expropriation of surpluses from Third World countries, especially because of the reliance of these countries on the export of primary products. But the late geographer Stephen Britton applies the theory to tourism, arguing that, given the unequal relationships in the world economy, the LAC countries can assume only a passive role in the development of international tourism.[8] There can be no better example of this passivity than the case of the Guatemalan Mayans, whose condition William Blum describes thus: 'It would be difficult to exaggerate the misery of the mainly Indian peasants and urban poor of Guatemala who make up three-quarters of the population of this beautiful land so favoured by American tourists.'[9] It is difficult to deny the fundamental point behind Blum's description that much of the tourist appeal of Guatemala is provided by its indigenous people – their colour, their culture and their activities – all of which mask their poverty and the oppression that they suffer.

After the Second World War, a slow opening up of the southern part of the continent to the growing middle classes of the northern half of the continent gradually turned into a veritable flood of tourists by the end of the century. First were the would-be explorers tracing the footsteps of earlier explorers and earning kudos for their trips when they returned with cultural and ethnic trophies for their walls and mantelpieces back home. Then the fieldworkers carrying out their own research accessed every nook and cranny of the region and returned with their results to claim their PhD or other form of academic standing. The early climbers searching for a more exotic challenge than the Alps found it in the Andes. All these groups, and others, reported back (mainly to Europe and North America) and enticed a new wave of travellers to venture further than they had previously done – the late 1960s and 1970s saw Latin America become a favoured destination for the early backpackers, representing the extension of hippy culture into the world of travel. They reported back with tales of wild journeys, exotic ecosystems and strange cultures, and in so doing gave rise to that particular branch of tourism known as backpacking, the experience of which is now almost a rite of passage for many young people around university age.

At around the same time, mass air travel was opening up beach resorts such as Cancún in Mexico (see Chapter 4) and Montego Bay in Jamaica. Whilst these were generally avoided by the backpackers, they gave access to a certain degree of exoticism to a growing number of people who did not want to get off the beaten track or experience the difficult and different circumstances so beloved of the back-packers. They did, however, want to have the sun, sand, sea and sex experience with a surety of good weather, but in the luxury of First World conditions rather than in the paucity and harshness of Third World conditions. Mass enclave tourism was born; and cruise ship tourism boomed too.[10]

At the end of the 1970s and throughout the 1980s, solidarity tours began to emerge with increasing awareness of liberation struggles in the subcontinent. The romanticism of Ernesto Ché Guevara and fascination for and sympathy with insurgent movements fighting for social justice was, and perhaps still is, a motivation for many more independent tourists and backpackers. Relative to the tourism industry as a whole, however, this group of tourists has remained minor in significance and such tours lost some impetus with the fall of the Soviet Union and the seeming helplessness of the struggle against the only remaining superpower, the USA. But recently the growing and maturing anti-globalisation movement has inspired something of a rebirth of this form of group visit – although still minor in significance – but in the slightly different guise of themed political tours such as Fair Trade tours, Effects of Globalisation tours, Human Rights tours, and so on.

Throughout the 1980s as global environmental concerns such as the destruction and rate of loss of tropical rainforests gained ground in the consciousness of the growing First World middle class, ecotourism leapt into prominence and was seized upon by some governments, such as Ecuador, Costa Rica, Belize and Dominica, as a chance to gain an advantage in the development of the tourism industry. Tourism for the purposes of seeing wildlife and marine life, experiencing tropical ecosystems and learning about indigenous cultures gained popularity with those who could pay for it. Ecolodges sprang up all over the continent; canopy walks gained a certain cachet for the ecotourist, the linking of research with tourism also grew, the two merging at times to become indistinguishable from each other. At the same time, national governments quickly perceived the need to show the tourists how they were looking after their environments by designating many areas as protected (see Chapter 4). In many cases this was purely for tourist consumption and was more of a paper exercise than a reality on the ground, but in others conflicts arose between local residents, tourism service providers and park or reserve rangers as a result of the area protection status. Area protection programmes, whether real or paper exercises, also demonstrated to national populations the national government's commitment to their own territory and helped entice some of the national population – those who could afford it – into the world of tourism as a receiver of the experience rather than as a service provider or a bystander. Co-option of local environmental groups by the government in these area protection measures was important in this respect. In Honduras, for example, many of the more than fifty major protected areas are managed by local environmental organisations under contract to the government.

At the start of the twenty-first century, the distinction between the First World and the Third World, or between the world of the 'haves' and the world of the 'have-nots' seemed to grow ever clearer and wider; and this prompted new features of the tourism trade such as the *favela* tour (of Rio de Janeiro for instance – see Chapter 6) and the linking of tourism development with the reduction of poverty (see Chapter 3). These examples come under the category of 'responsible tourism' which according to some commentators is gaining popularity over its competitor 'ecotourism'. The environmental commentator John Vidal suggests that ecotourism has become somewhat confused because of its association with 'greenwash' (see Chapter 4) whilst 'the demand for responsible tourism is thought to be growing strongly'.[11]

But the finer points of tourism terminology are not an issue of great concern here. Responsible tourism, ecotourism, environmentally friendly tourism, sustainable tourism, development-related tourism, solidarity tourism, wildlife tourism, nature tourism, and many others – even including postmodern tourism, whatever that may be – are all terms found in tourism brochures and travel reviews and magazines at present. Essentially, their use reflects an increasing awareness of both the positive and negative impacts of tourism, an increasing desire to be seen to be responsible in one's travel and holiday behaviour and an increasing need to dissociate oneself from the masses. The latest editions of the *South American Handbook* (2004) are now aimed much more clearly at the independent traveller rather than the businessmen of earlier years and reflect the myriad opportunities for tourist activities of all types. In fact, all the guide books now appear to claim adherence to the principles of responsible tourism.

This increase in awareness of the roles and significance of tourism was also reflected in the United Nations declaration of 2002 as the International Year of Ecotourism, the World Tourism Organisation (WTO/OMT) becoming a fully fledged United Nations Agency in 2003 and the recognition given to the tourism industry in the General Agreement on Trade in Services (GATS) promoted and administered by the World Trade Organisation (WTO/OMC).[12] These events acknowledge the role of the tourism industry as a generator of economic growth and development and as an important plank supporting the prevailing model of neoliberal economic development.

Conversely, it might be suggested that such recognition by or for such august bodies merely acknowledges the significance of the role of the tourism industry as a potential vehicle for continued economic exploitation of the LAC countries. Visitors from the so-called First World now extract not only the wealth of minerals and resources from the LAC countries but also the gold of tourism – that is, the natural world of protected areas and the variety of cultural experiences. As Brazilian Severino Resende notes, the Third World might be 'transformed into a botanical garden for the First World, in the same way that [it was] transformed into breadbaskets of the former colonial empires'.[13] History repeats itself as Figure 2.1 suggests. In return, the tourists give little in the way of finance (as most of the moneys paid are repatriated to parent companies in the First World), the local cultures become westernised (for good or bad), and local pollution associated particularly with western consumer lifestyles increases.

Increasingly, though, this tide of First World tourists to the LAC countries is being amplified by those sectors of the LAC national populations who can afford to enjoy the tourist experience. Krishna Ghimire provides some interesting case studies to illustrate this rise in national tourism in his book *The Native Tourist*.[14]

Figure 2.1 History repeats itself . . .

Source: Adapted by Brian Rogers from a cartoon by the late Hugo Diaz, a Costa Rican cartoonist.

This thumbnail sketch of the intertwined history of the LAC countries and the tourism industry leads us into the industry's current role as either agent of exploitation or agent of development, sustainability and responsibility, or both, and we examine this debate in the following section.

The notion of development and the Millennium Development Goals

One of the major goals of responsible tourism is that it should contribute towards development. But to determine whether it does in fact act as an agent of development, a means through which development can be achieved, or is simply a new industry which allows the continued exploitation of the wealth, resources and people of the LAC countries by foreign entrepreneurs and their companies, it is necessary first to examine the notion of development. Is it something which is wholly benign? Is there only one channel that nations can follow in order to reach a developed stage or state? Who defines related notions such as underdevelopment? How are these notions defined? And how are tourism and development related?

Most theories of development have emerged from eurocentric thinking and analysis of western capitalist economic history, and the economic responses to development are derived from First World governments and First World-dominated multilateral institutions which apply a 'one-size-fits-all' paradigm of development. As Hettne contends, 'Once the first industrial nation had been born it provided the model to imitate . . . Not to imitate would mean permanent dependence . . . In order to develop it was deemed necessary for the "new nations" to imitate the Western model.'[15] Sachs maintains that since the day when US President Harry Truman declared large parts of the world to be 'under-developed', development, above all else, has signalled the need to escape the undignified confines of underdevelopment.[16]

Currently, this is often referred to as 'the only game in town',[17] which according to Rist is based on 'a concept which is supposed to command universal acceptance but which – as many have doubtless forgotten – was constructed within a particular history and culture'.[18] The concept is economic growth, on which western capitalist development is entirely dependent, and which has become *the* paradigm that has dominated and remained virtually unchallenged throughout the age of industrial and technological development. Amongst the supranational organisations, governmental institutions and commercial sectors of the industrial nations there is a consensus that development is contingent upon economic growth and wealth creation; and there is an almost universal acceptance that 'efficient markets are indispensable for effective development'[19] and that growth is dependent on a 'continuation of market-based policies'.[20] Yao Graham refers to this as 'the monoculture of a single development model rooted in neoliberal economics. Its suitability for all is taken for granted, as is the intrinsic good of market forces and trade liberalisation.'[21] This leaves the primacy of economic growth unquestioned. But Rist and many others suggest that we must question this notion of development as the only one available. He argues that we must face the possibility that the western capitalist model of development, far from 'bridging the ritually deplored gap' between rich and poor nations, 'continues to widen it'.[22]

Rist's point reflects the rise to greater prominence since the early 1990s of the need to reduce levels of absolute poverty in Third World countries, not least because of the growing chorus of criticism from the anti-globalisation movement. Escobar summarises the major points of this criticism thus:

> Instead of the kingdom of abundance promised by theorists and politicians in the 1950s, the discourse and strategy of development produced its opposite: massive underdevelopment and impoverishment, untold exploitation and famine. The debt crisis, the Sahelian famine, increasing poverty, malnutrition and violence are only the most pathetic signs of the failure of forty years of development.[23]

As Eric Hershberg has described it, 'the evidence of failure now exceeds the capacity for denial of even the most impervious economists'.[24]

The response to these critics has varied according to the institutional rhetoric accorded to the acknowledgement of poverty and inequity as major problems, the availability of suitable scapegoats such as dictators or corruption, and the perceived need to blame or absolve from blame the prevailing model of development. One response of world leaders in 2000 at the United Nations Millennium Summit was to agree to a set of time-bound and measurable goals and targets for combating poverty, hunger, disease, illiteracy, environmental degradation and discrimination against women. These are now referred to as the Millennium Development Goals and were set to be achieved by the year 2015. The goals and targets are shown in Table 2.1. For each target a number of indicators are used to measure the success or failure of their achievement.

A crucial point about the response to critics of capitalist economic development, however, and about the Millennium Development Goals has been the unwillingness or inability of the controlling institutions and governments to deviate from the ideology of economic growth as the engine for development. The response of the World Bank for instance is partly characterised by its annual World Development Report in which it addresses the problems of poverty and inequity and which it uses to portray its work as 'development'. But in so doing it describes development largely in terms of growth and avoids addressing the contradictions between growth-led economic development and increases in poverty and inequality. Rather than deviate from the ideology of growth, the UN agencies have begun in earnest their exhortations to national governments of the industrialised, technocratic countries and international financial institutions to adopt policies in order to meet the Millennium Development Goals – to follow, in other words, what might be called a reformist agenda. This does not mean that the UN is advocating a change in ideology; rather, it is simply trying to fine-tune the existing model in order to mitigate its worst effects and to excuse its own role in the creation of poverty and inequality through the global application of western capitalist economic development.

Table 2.1 The UN's Millennium Development Goals

GOAL 1   Eradicate extreme poverty and hunger
Target 1   Halve, between 1990 and 2015, the proportion of people whose income is less than one dollar a day
Target 2   Halve, between 1990 and 2015, the proportion of people who suffer from hunger

GOAL 2   Achieve universal primary education
Target 3   Ensure that, by 2015, children everywhere, boys and girls alike, will be able to complete a full course of primary schooling

GOAL 3   Promote gender equality and empower women
Target 4   Eliminate gender disparity in primary and secondary education, preferably by 2005, and to all levels of education no later than 2015

GOAL 4   Reduce child mortality
Target 5   Reduce by two thirds, between 1990 and 2015, the under-five mortality rate

GOAL 5   Improve maternal health
Target 6   Reduce by three quarters, between 1990 and 2015, the maternal mortality ratio

GOAL 6   Combat HIV/AIDS, malaria and other diseases
Target 7   Have halted by 2015 and begun to reverse the spread of HIV/AIDS
Target 8   Have halted by 2015 and begun to reverse the incidence of malaria and other major diseases

GOAL 7   Ensure environmental sustainability
Target 9   Integrate the principles of sustainable development into country policies and programmes and reverse the loss of environmental resources
Target 10   Halve, by 2015, the proportion of people without sustainable access to safe drinking water
Target 11   By 2020, to have achieved a significant improvement in the lives of at least 100 million slum dwellers

GOAL 8   Develop a global partnership for development
Target 12   Develop further an open, rule-based, predictable, non-discriminatory trading and financial system
Target 13   Address the special needs of the least developed countries
Target 14   Address the special needs of landlocked countries and small island developing states
Target 15   Deal comprehensively with the debt problems of developing countries through national and international measures in order to make debt sustainable in the long term
Target 16   In cooperation with the developing countries, develop and implement strategies for decent and productive work for youth
Target 17   In cooperation with pharmaceutical companies, provide access to affordable, essential drugs in developing countries
Target 18   In cooperation with the private sector, make available the benefits of new technologies, especially information and communications

---

The most recent efforts to alter the policies of what James Petras calls the 'Golden Age of Imperial Pillage (1975–2000)'[25] are those of several Latin American electorates who with their votes have indicated a general distaste for the ever-widening gap between the rich and the poor and have elected governments less devoted to the extreme orthodoxy of neoliberalism. These include Venezuela and Bolivia particularly – or as President Bush might call them, 'The Axis of Not-Quite-So-Evil' – but to a lesser degree Argentina, Uruguay, Brazil, Ecuador, Peru, Chile, Panama and Nicaragua. It is still a subject of considerable debate, however, whether the Latin American trend really represents a departure from the prevailing patterns of trade and the prevailing model of development or just a reordering of priorities within the same model.

In the contemporary development agenda the power, prestige and wealth of First World economies is non-negotiable. This has relevance to the position of the international tourist visiting the LAC countries on whom is reflected this power and prestige. Many of the LAC countries are often characterised by their high levels of poverty and inequality. Tourism is being increasingly invoked as an agent of development, a means through which development can be achieved. This is witnessed by the recent efforts of First World governmental overseas development agencies and the supranational organisations such as the WTO/OMC and the United Nations Development Programme (UNDP) as well as the WTO/OMT to promote pro-poor tourism initiatives, a relatively new feature of the industry which we examine at greater length in Chapter 3. These initiatives arise from within First World governments and the supranational agencies of western capitalism and are predicated on western capitalist expansion and development.

The Millennium Development Goals themselves make it clear that they are to be achieved, or not, in collaboration with the private sector, and the former UN Secretary General Kofi Annan talked of a 'successful, development-oriented result [which] could boost investment flows and help revive the global economy'.[26] Indeed, it is interesting to note that all the rhetoric about poverty reduction and elimination that has emanated from the supranational institutions in recent years has assumed that there will be and can be no change in the prevailing model of development. The point is not that the intentions to adjust policy are ill-inspired, but that they are contingent upon a system which has so manifestly failed to deliver development to a majority of the world's population.

Of course, not everyone would see such a system as being a failure – it depends on whose interests it represents – and some would dispute that a majority of the world's population have not benefited from half a century of 'development'. But if development is intended as something to benefit everyone, as the UN clearly aims, then increasing levels of poverty and inequality in LAC countries would seem to be a measure of failure.

In 2003, Gordon Brown, the UK Chancellor of the Exchequer, proposed the establishment of an International Financial Facility (IFF) to attract the US$50 billion a year required to meet the UN's Millennium Development Goals. (It is worth noting that Kofi Annan, the UN Secretary General, states that 'US$100 billion a year is universally recognised as necessary to achieve the millennium development goals'.[27]) Along with James Wolfensohn, then President of the World Bank, Brown launched a defence of the proposal in early 2004 on the grounds that by 2005 the first goal, for girls' education, will go unmet, and they stated that world leaders face the stark choice: 'To have a chance of meeting the millennium goals, a new deal between developing and developed countries must be forged. It is in the interests of developing countries to tackle corruption and undertake a sequenced opening up to the investment, trade and growth that will provide jobs.'[28] This continues the prescription of liberalisation offered to all under-developed countries – opening up national markets for foreign investment which implies that tourism developments will be encouraged – a prescription which, as we shall see in the following section, has served as something of a con trick to enable the continued dominance of First World corporations in pursuit of profits. Some critics fear that the sustainable development of which Brown and Wolfensohn talk so frequently, 'might be about making up for the wasteful lifestyle for the rich by proposing a frugal one for the poor'.[29]

In 2005 the UK held the presidency of the G8 countries[30] and promised to make the raising of the required US$50 billion a year a priority. As George Gelber, head of policy at the Catholic aid agency CAFOD, warned: 'If the eight most powerful world leaders fail to come up with the $50bn we need, they might as well tear up the solemn millennium pledge that they made in 2000.'[31] In early 2005, however, the signs were not good, as the US government poured cold water on Brown's IFF initiative and the G8 governments were divided by alternative proposals from Germany (for an air fuel tax designed to help the poor) and France (for a currency speculation tax for the same purpose).

Many of the UK-based international NGOs would also pour cold water over such ideas – for different reasons – on the grounds that proposals such as the IFF and others stemming from the supranational institutions and G8 governments are more likely to form part of the problem rather than part of the solution to the failure of development.[32] The final collapse of the Doha round of trade talks in summer 2006 served as an admission of defeat for the agents and proponents of western capitalist economic development in their efforts to portray themselves as serving the interests of the poor and disadvantaged of the world.

It is notable that this widely acknowledged collapse of the trade talks occurred at more or less the same time as a meeting of representatives from Latin American, African, European and Central American integration parliaments held in Guatemala in June 2006. At the meeting, the President of the Central American Parliament, Julio Palacios, declared that, six years after the signing of the Millennium Goals by the United Nations, there were very few achievements regarding poverty reduction to show. 'Our childhood is shipwrecked to an uncertain future; the distribution of wealth turns more unequal every day; unemployment is a constant problem for families,' said Palacios in his opening speech at the World Parliamentary Summit. He also pointed out that 'Cuba is the only country in the Caribbean and in Latin America that surpassed almost all areas of the Millennium Goals'.[33]

Despite that failure, in their attempts to bridge the inequality and poverty gap through development, the proponents of development have begun to promote tourism. Development is important to the tourist and the tourism industry. On the one hand, many visitors from the First World travel to the LAC countries precisely because they are defined as 'under-developed'; they want to see and experience development at a stage that is different from that of their own advanced nation – partly because there is kudos in such an experience for them when they return to their home environment and social milieu. (Of course other motives can also be found.) On the other hand, under-development is also something of a deterrent to the tourist. The majority of tourists don't travel to the LAC countries to be poor; but some of them travel there to see poor. The industry therefore has to provide facilities which are developed rather than under-developed, and this necessitates the intervention of First World supranational organisations, such as the World Bank and IMF, and the inward investment promoted by international private banks and TNCs, to provide the funding to install the physical and economic infrastructural development that the tourists and tourism industry depend on.

Tourism and free trade

Free trade and economic growth are regarded as essential prerequisites of develop-ment by the supranational and national agencies: the World Bank, IMF, OECD, EU and G8 governments. The discourse of these agencies is commonly referred to as the Washington Consensus and its ideology of free trade is captured in the term neoliberalism. The ideology of free trade is synonymous with economic globalisation, and the spread of capitalist relations of production is pervasive and is witnessed in the rapid spread of tourism. We address the linkages between free trade, tourism and responsibility in this section.

Neoliberalism sees development as a single linear progression of economic growth and wellbeing, a model that has been portrayed by a number of theorists as a series of steps or stages (such as Rostow's *The Stages of Economic Growth*[34]) to the 'promised land' – a developed state. Similar conceptual models (such as Butler's Tourist Area Cycle of Evolution[35]) have been advanced in tourism circles. It is an almost universal acceptance that there can be no 'development' without economic growth and no economic growth without free trade, an equation that commands near universal respect but which Rist and others insist needs to be questioned.[36]

Free trade fundamentally requires open and deregulated markets which provide a level playing field to those who operate business such that there exist no restrictions to trade. As the WTO/OMT explains, 'In order to do business as effectively as possible, companies need level playing fields so that they can have equal access to natural resources, expertise, technologies and investment, both within countries and across borders.'[37] In theory this would allow national economies to specialise in producing goods or providing services that they are best at producing or providing. They would export these goods or services and import the things that they are not so good at producing. This is the notion of comparative advantage, one of the theoretical planks on which the ideology of free trade is built and which can work very much in the favour of the tourism industry in countries which have the comparative advantage of spectacular environmental and cultural attractions.

The notion of free trade itself, however, manages to hide the uneven and unequal economic interdependencies manifest for example in trade protectionist measures adopted by First World governments. In this regard the hegemony of the United States of America is noteworthy, though equally the power vested in the WTO/OMC, the G8 and key trading blocs (especially the North American Free Trade Agreement and the European Union) should not be underestimated. Arthur MacEwan challenges these orthodoxies arguing that neoliberalism is:

> based on the ideology of free markets, and the practitioners of neoliberalism argue that free markets are the best means by which to support the interests of US businesses and, through US business, the interests of the US economy and US citizens. Yet guided by the fundamental principle that their goal is to promote US business interests, US policy makers must continually violate free market principles.[38]

The theoretical level playing field belies the inequalities of development, economy and power which exist between and within the First World and the Third World. In the case of the LAC countries, Wooding and Moseley-Williams tell us that 'before the current phase of globalisation, [this] was the region of the world where the greatest inequalities were found', and that 'today it is even more unequal'.[39]

Fred Rosen is also critical of the theoretical framework of the Washington Consensus, arguing that it exists 'at the height of abstraction, de-linked from reality',[40] and Wayne Ellwood suggests that 'Talk of "level playing fields" and "pure competition" obscures the evidence that poor countries are severely disadvantaged to begin with'.[41] In existing circumstances, it can be forcefully argued that there can be no such thing as free trade or a level playing field.

The deregulation of the local business environment – deregulation is another of the theoretical planks on which the ideology of free trade is built – to create the hypothetical level playing field actually creates the conditions which clearly favour those with access to wealth and resources. Stealing Bernard Crick's phrase, we might describe deregulation as 'the dramatic bonfire of the controls needed to make free markets tolerable'.[42] This access to the global infrastructures of transport and communications on which free trade and the tourism industry depend is itself unevenly distributed. People and businesses in the First World can easily reach and make use of resources and cultures in the LAC countries whilst a substantial majority of the national populations of the subcontinent do not enjoy such ease of access despite local infrastructural developments. To some extent this variability in the ease of access to modern business infrastructure is illustrated by the example of the Plan Puebla-Panamá given in Box 2.1.

Clearly, for the international tourism industry, improvements in airport links, electronic ticketing, banking facilities and other infrastructure are seen as helpful and promotional. New roads enable the coaster buses to transport tourists to ever

Box 2.1   The Plan Puebla-Panamá (PPP)

The Plan Puebla-Panamá (PPP) is a co-ordinating plan for infrastructural development projects throughout the isthmus of Central America and exemplifies the unevenness of the benefits of infrastructural developments made in the name of free trade. The Plan was prepared and proposed in 2000 by Mexican President Vicente Fox and is intended as a 25-year, $20 billion programme, one of whose major components is tourism promotion. The Plan calls for major new or improved transport arteries, new power lines to distribute electricity generated by gas and water which will require the construction of numerous new dams, six massive 'development zones' for *maquiladora* plants and processing factories, the construction of at least two deep water ports with large-scale container handling facilities, oil pipelines and large-scale industrial development throughout the region.

Critics of the 'what-have-the-Romans-ever-done-for-us?' argument would point out that such developments will be there for all to use. But all the new transport, oil and power lines are viewed as major arterial developments, designed to facilitate the movement of goods from East Asia and South America to Europe and North America. As such, they will improve the turn-round time of capital profit to be made by transnational corporations – i.e., the speed at which profit can be made. Such tightly focused arterial development will do nothing for rural development except create conditions which will force people to migrate from the rural areas towards the possibility of low-paid, unsafe and unhealthy work in the new *maquilas*. Its emphasis on the privatisation of land and property is likely to dispossess indigenous communities of commonly held land over which they have no title deeds but which has belonged to their communities in some cases for centuries. Communities which are already marginalised and impoverished are likely to find their conditions for living worsened.

There has been no serious consultation with communities which will be affected by the PPP, and public information about it is scattered, incomplete and confusing. Moreover, few of the PPP-related projects call for environmental impact assessments. Reporting for Reuters in October 2002, Greg Brosnan wrote that indigenous groups in southern Mexico brought cross-border traffic to a standstill to protest against the PPP. They 'say no-one has consulted them on a plan they believe will benefit only big business. They fear it will endanger their livelihoods and destroy their culture, forcing them off scarce land and turning them into a source of cheap labour for foreign investors.' Brosnan quotes Crisanto Jimenez, a Mam Indian, as saying 'They want to build six-lane highways. They will destroy our lands and make us poorer.'[1] Indigenous groups in Costa Rica, Nicaragua and Honduras have also held protests against the free trade plan, and in El Salvador thousands of campesinos, students and workers blocked key transit routes in protest at the Plan.

Despite President Fox's claims that the PPP will also help promote tourism, health care and education in the region, its budget priorities are dramatically skewed. For instance, in the Mexican government's 2002 budget for the programme, 82 per cent of funding was for transport projects and less than 3 per cent was for health and social development projects with no specific attention to rural development. It is also claimed by its proponents that it will promote social development through private investment that will create jobs and thereby eradicate poverty. A CIEPAC[2] report from October 2002 views this notion as

> an absurd simplification. Neither public nor private investment automatically leads to higher living standards for the poor, unless steps have been taken beforehand to eliminate the structural injustices that exist in the economic, political, social and cultural spheres. In fact, investment often deepens poverty, as has been the case during the last 20 years of neoliberal policies, precisely because existing injustices have not been eliminated. Thus the rich and powerful benefit more from investments . . . The plans and projects are designed in collaboration with and for big business, not for the 65 million people who live in the PPP area, the vast majority of whom are in poverty (75 per cent living with less than US$2 a day).[3]

Sources and notes:
1   Greg Brosnan (2002) 'Central American Indians Protest Development Plan', Reuters report, 15 October 2002.
2   CIEPAC: The Centre for Economic and Political Investigations of Community Action.
3   CIEPAC (2002) 'Plan Puebla Panama Primer', www.ciepac.org, 21 October 2002.

more remote environments. A potable water supply is for many a *sine qua non* for making a visit, and an advanced sewerage system is deemed by many tourists from advanced countries as the most important of requirements. In many LAC environments neither running water nor sewerage systems are available. In some cases, especially in the Caribbean, this has given rise to the all-inclusive resort, built with its own water purification plant and sewerage treatment or disposal system, surrounded by communities which do not have the same trappings of development. The Costa Rican cartoon in Figure 2.2 illustrates these differences.

Many of the infrastructural developments in the LAC countries that we are referring to here have been built over the last two or three decades with loans from the supranational organisations listed above, especially the World Bank and the Inter-American Development Bank (IDB). Much of the work for these projects was

Figure 2.2 All-inclusives and the local environment

Source: Adapted by Martin Mowforth from a Diaz cartoon in Costa Rican Semanario Universidad, 13 February 1992.

carried out by transnational companies (TNCs) based in the First World or by national subsidiary companies set up by these TNCs. As Martin Khor explains, the modernisation that such infrastructural development represents 'would create conditions whereby the Third World would become dependent on the transnational companies for technology and inputs'.[43]

It is these loans, and rising interest rates on them, which Third World countries accepted which created the so-called debt crisis of the 1980s. The debt has not disappeared, nor has the crisis (as witnessed by the precipitous Argentinian economic collapse of 2001), but the term 'crisis' is used less often to describe what is now seen almost as a 'normal' condition. And it is the debt condition which enables the supranational organisations to dictate the economic policies to be pursued by the LAC countries. Not too surprisingly, the supranational organisations dictate the pursuance of a set of policies which promote the notion of free trade, the neoliberal agenda, from which the TNCs of the First World will be the major beneficiaries.

This set of policies has three broad strategies for any given country: to increase earnings of foreign capital; to reduce state involvement in the economy; and to use fiscal measures to regulate the level of economic activity and the balance of payments. Box 2.2 briefly outlines how these three policy planks are carried out in most LAC countries and their significance to the development of tourism.

Box 2.2   Policies and conditions associated with structural adjustment programmes/Poverty Reduction Strategy Papers

A   Increase earnings of foreign capital

*   *Boost export production*
    One implication of this is that the production of goods for local and national needs is relegated in importance as a result of the drive for export production. Tourism is seen as an export product, attracting foreign currency into the country. Many Third World countries which used to be self-sufficient in foodstuffs now have to import basic grains and other foodstuffs. In 1996, for instance, Honduras began to import beans for the first time in its history.

*   *Devalue the national currency*
    Devaluation makes the country more attractive to foreign investors seeking a cheap place to produce their goods. The cheapness of its currency against other currencies also increases its attraction to foreign tourists.

*   *Reduce and/or abolish import tariffs*
    Import tariffs are designed to protect the price of the production of goods for home consumption against the entry of cheaper imported goods. In practice, removing them can have the effect of flooding the country with expensive luxury goods and durables with high value added in production from the First World – for purchase by tourists and the rich élite.

B   Reduce state involvement in the economy

*   *Privatise state-run enterprises*
    Tourism is a flagship of the private sector.

*   *Cut public spending on activities which cannot be privatised*
    For example, social services, health provision, education – although increasingly attempts are being made to privatise these services too.

*   *Deregulation*
    Reduction of all forms of controls, especially economic regulations, which inhibit industry's ability to maximise profits.

C   Fiscal measures

*   *Reduce inflation*
    This increases security of investment, including investments in tourist developments.

*   *Cut interest rates*
    This increases investment, including investment in the tourism industry.

*   *Encourage investment rather than saving*

If all companies (whether in the tourism industry or any other industry) were of the same size and scale, with the same advantages and disadvantages and the same resources, then clearly the 'level playing field' would offer fairness. As it happens, however, companies from the First World which are involved in the tourism industry have a number of advantages over their counterparts in the LAC countries. They have access to financial, electronic, information and communications systems and technology that are the product of many decades of investment, research and development – in fact a history of industrial and technological development since the early days of the Industrial Revolution. They have legal, governmental and establishment procedures which have had a long time to evolve and knowledge of which is essential for business success in the international market place. And they have a political lobbying power that reflects their financial resources, which in some cases are greater than the governments with which they enter into negotiations.

The strategies briefly described above, then, will favour First World companies over nascent local companies in the LAC countries and, as tourism plays an important role in the free trade treaties currently under negotiation between different groupings of the LAC countries and the USA and the European Union, there are credible concerns that local tourism companies will lose out to US and European operators entering into their market. In recent years many local companies have had to enter into uneven negotiations to work as 'partners' with First World TNCs which dictate the terms of business. This relationship between First World TNCs and national companies in the LAC countries is particularly pertinent to the tourism industry. It would be possible, for instance, for a tourist to depart from Europe in a plane of the Spanish airline Iberia, be transferred from the LAC country airport in a Volkswagen minibus (made in Germany) to a hotel belonging to the Spanish hotel TNC, Grupo Barcelo. They could eat at Spanish, Italian, French or British owned restaurants whilst reading their guide book written by a fellow European. They could shop in a mall where the majority of shops are owned by either North American or European TNCs and would return home in a plane of the same airline company, the whole tour having been arranged for them by a European tour company. Then they would be able to tell their friends and family what a marvellous place the LAC countries are. The significance of this example, as Paul Gonsalves points out is that: 'ownership, control and therefore benefits, from Third World tourism, accrue mainly to the rich industrialised nations from where the tourists originate'.[44] Ensuring that a significant proportion of the holidaymakers' payments finds its way into the local economy and into local hands is often seen as an important element of responsibility in tourism. If that is so, then the example just given is clearly irresponsible.

In the tourism industry under a deregulated and privatised system of conducting business and faced with competition from companies with the advantages listed above, local companies can do little but submit to takeover bids and arrangements with foreign parent companies. They have the advantages of local knowledge and local contacts, and these they can use as bargaining points to improve their standing in the franchise agreements they have to accept from the invading company from the USA or the EU or south-east Asia. But these are nothing in comparison with the power of capital and technology available to the First World TNC. The TNC knows that the majority of its clients – their profit – are either from the First World or from middle-class and elite sectors of Third World societies and require First World conditions for their visit. Crucially, the TNCs operate under a system of profit maximisation in which all costs other than those which are financial are externalised – or, in other words, ignored. These external costs of tourism operations include costs borne by the local environment, local people, local society, local economy and local culture where they operate. As David Korten states, 'Increasingly, it is the corporate interest more than the human interest that defines the policy agendas of states and international bodies'.[45] As Deborah McLaren remarks, 'Nowhere is this more true than within travel and tourism'.[46]

Taken together, deregulation and privatisation are often referred to as liberal-isation, the economics of which 'are based on the idea that tourism will yield overwhelmingly positive benefits on growth linked with development'.[47] In their own words, the WTO/OMT 'advocates progressive liberalisation, . . . "liberalisation with a human face" [to] encourage the rapid expansion of tourism in line with market capacity so that it can serve as a dynamic catalyst of employment, wealth, investment and poverty elimination'.[48] In its liberalisation campaign, the WTO/ OMT has vowed to work closely with the WTO/OMC to emphasise 'the importance of open markets for socio-economic development', 'the need to eliminate all barriers to tourism growth', 'the need to ensure that liberalisation creates fair conditions of competition ensuring a level playing field for growth of all countries' tourism services', and 'the vital reliance of tourism on free flowing transport systems', amongst others.[49]

But as Williams argues, 'This ignores distributional impacts and forward and backward linkages which are endemic problems in developing countries' economies . . . While tourism may bring employment, it is often seasonal and highly exploitative.'[50] And Larry Elliott makes the point that 'the brutal truth is that many poor countries are in no shape to take advantage of more open markets, even assuming they become available. Moreover, if trade does increase, the benefits will be felt by a small privileged minority and not by the bulk of the population.'[51] Even the US Central Intelligence Agency (CIA), not known for its affiliation to the anti-globalisation movement, warns that:

> The rising tide of the global economy will . . . spawn conflicts at home and abroad, ensuring an ever wider gap between regional winners and losers than exists today. [Globalisation's] evolution will be rocky, marked by chronic financial volatility and a widening economic divide. Regions, countries and groups feeling left behind will face deepening economic stagnation, political instability, and cultural alienation. They will foster political, ethnic, ideological and religious extremism, along with the violence that often accompanies it.[52]

Insofar as the international tourism industry is strongly associated with the evolution of the global economy, the CIA's warning is relevant here. Nobody needs to be reminded of the effects on international tourism of the attacks on the USA of 11 September 2001.

Vandana Shiva suggests that the neoliberal agenda pursued by the centralised and undemocratic WTO/OMC is 'driven by the objectives of establishing corporate control over every dimension of our lives' and she argues that control by the TNCs should give way to the 'rights of all species and the rights of all people . . . before the rights of corporations to make limitless profits through limitless destruction'.[53] The cartoon in Figure 2.3 perhaps overstates this corporate exploitation of local environments and societies by elements of the tourism industry, but is illustrative nevertheless.

It is clear from the WTO/OMT's description that liberalisation of the tourism industry is seen as a general development strategy. Relating this to a process of industrialisation, Joe Bandy argues that ecotourism is similar to the export-oriented development model insofar as foreign investment for luxury services is encouraged through ecotourism and distorts local investment and exploits divisions of labour.[54] Stretching the analogy, the nature tourism site is seen as similar to export processing zones for commodities, the notorious sweatshop work conditions being replaced by general subservience of the local populace to the wealthy visiting tourists. The most recent round of global economic liberalisation, which includes particularly the tourism industry, is driven by the General Agreement on Trade in Services (GATS) whose significance to the tourism industry is explored further in the following section.

Tourism and the General Agreement on Trade in Services (GATS)

The GATS is one of several trade agreements administered and controlled by the WTO/OMC. It represents the extension of economic liberalisation measures (such as privatisation, deregulation and reduction of trade barriers) from primary and secondary sectors (involving the production of goods) into the tertiary and quaternary sectors (involving trade in services, including tourism). The principal motive for the most recent round of economic globalisation, of which GATS is the newest agreement, is arguably to ensure the continued expansion of markets for services provided from the industrialised nations, states in which the service sector has already been heavily commercialised and is dominated by TNCs.

In practice, according to a report by the World Development Movement (WDM), one of the major UK critics of GATS, 'WTO staff and negotiators openly acknowledge that GATS exists only because of pressure from service multinationals and that this influence has continued since GATS came into effect'.[55] Moreover, the European Commission's website describes GATS as 'not just something that exists between governments. It is first and foremost an instrument for the benefit of business.'[56]

In theory, the WTO/OMC works on a consensus principle to create a rules-based system for the conduct of multilateral trade, and GATS works on the same principle for the same aim. The mechanism of reaching agreement by consensus appears to

Figure 2.3 Globalisation and the tourism industry

Source: Adapted by Brian Rogers and Martin Mowforth from a cartoon by El Fisgón.

be given as a major justification for the GATS. But whether a genuine consensus was achieved at the Doha Ministerial meeting in Qatar in November 2001, called to discuss GATS, or at previous and later similar meetings, is highly debatable: for instance, 481 delegates representing the G7[57] nations were present at Doha whilst 276 delegates represented the 39 Least Developed countries, and it is difficult if not impossible to achieve consensus agreement in such unbalanced circumstances. Indeed, the WDM also reports that widespread pressure was exerted by industrialised countries on Third World country negotiators, including suggestions that aid would be withdrawn.[58] In general, the GATS appears to reflect and reinforce rather than challenge the existing unevenness and inequality in the global economic system.

Under the agreement, any signatory country has to accord the same treatment to foreign companies as it does to domestic companies in specified sectors of trade to which they have committed themselves. As Kalisch explains, it

> is designed to ensure that host governments, confronted with powerful transnational corporations who import both their own staff and the majority of goods needed for their tourism operation, cannot compel them to use local materials and products to enhance the 'multiplier effect', or to take special measures to secure a competitive base for their domestic businesses.[59]

The GATS defines import tariffs, subsidies and other measures designed to assist domestic companies as 'trade-restrictive', and thereby effectively undermines the power of governments to legislate in the national interest. For instance, the US hotel lobby has submitted a list of 'regulatory obstacles' used by other countries that it wants eliminated through new WTO/OMC rules. Many of these trade-restrictive obstacles are the policies needed and used to ensure that local communities benefit from tourism. An example of this is the European Union's request that Mexico should totally eliminate the need to obtain a licence to open a hotel, restaurant, tour operation or tour guide business in Mexico. The licence, however, is an important means by which the Mexican government can regulate the tourism sector.[60]

In GATS, tourism falls into the sector labelled *Tourism and Travel-Related Services* which has four categories: hotels and restaurants; travel agencies and tour operator services; tourist guide services; other (unspecified). Kalisch points out that under GATS:

> it would be impossible for a government to impose on foreign companies practices to limit negative environmental, social and cultural impacts in their country, such as restricting the mushrooming of foreign owned development, (including all-inclusive hotel developments, which are highly controversial among local people because they contribute almost nothing to the local economy), and making the employment of local workers, the use of local products and materials a condition of their investment. It could also render a government powerless to stop tourism developments on indigenous land, including sacred sites, in response to community protests.[61]

Kalisch's analysis also suggests that GATS is likely to place fair trade in tourism – see the following section – at a disadvantage since 'the beneficiaries of Fair Trade in Tourism are poor communities, for whom the informal sector (vendors, traders, guides) is an important source of income, as well as small entrepreneurs and workers employed in the tourism industry'. Representing the UK NGO Tourism Concern, Kalisch identifies the following concerns about the role that GATS is likely to play in the emerging tourism industries in Third World countries:

- The manner in which GATS negotiations are carried out in the WTO/OMC disadvantages and marginalises Third World countries and civil society organisations.
- GATS fails to integrate sustainable development as identified in major international treaties and agreements, such as the Convention on Biological Diversity which includes decisions on tourism.
- GATS does not address the specific environmental, social, economic and cultural impacts of tourism in a destination.
- Expanding global tourism increases the threat of biopiracy under the guise of ecotourism, when tour operators and clients collect rare medicinal plants from areas inhabited by tribal peoples but without consulting them.
- GATS makes no provision for the monitoring and regulation of TNCs.[62]

As K. T. Suresh, of the NGO Equations, suggests, it is the wrong framework, as it may prevent the necessary regulation of the tourism industry.[63]

Of all these concerns of relevance to the activity of tourism, it is perhaps the fear that the GATS will further deepen the increasing poverty and inequalities experienced in some Third World countries that prompts the greatest protest. In the last two decades of the twentieth century, structural adjustment was the major economic policy employed in up to ninety Third World countries under programmes imposed by the IMF and World Bank. During this time, according to Barry Coates, former director of the WDM, 'growth rates in Latin America have stagnated while inequality has risen dramatically'.[64]

The possibility that the GATS will further deepen these effects is feared not just by anti-globalisation protestors but also by the protagonists of fair trade in tourism and pro-poor tourism. It is clear from much of the propaganda of the supranational agencies which represent neoliberal economic development that they too have come to recognise the fact that neoliberal policy prescriptions have not solved the problems of poverty and inequality. Hence the recent rise of pro-poor tourism and poverty elimination or reduction programmes in their policy and mission statements.

Whilst we consider that criticism of GATS and other international free trade agreements is wholly justified and appropriate, it is worth adding that environmental and social protections in some LAC countries are already extremely weak and 'liberal'. For instance, there are few national constraints on corporations and property developers based in Mexico to prevent them from developing golf resorts in the Yucatán peninsula. Nevertheless, it should be emphasised that it is the express intent of GATS and international free trade treaties to reduce and eliminate the barriers. Their choice of the word 'barrier' is significant – to others, these barriers are 'protections'.

In the final section of this chapter we examine the type of measure that is meant to mitigate these ill-effects of free trade policies – measures such as voluntary self-regulation, codes of conduct, corporate social responsibility, and fair trade in tourism, all of which, along with others, are used, genuinely or disingenuously, in order to make tourism work to the benefit of those for whom free trade leads to deeper poverty and greater misery. These policy measures also include pro-poor tourism programmes and eco-labelling and certification schemes, but these are covered at more appropriate places in the book, respectively in Chapters 3 and 4.

A guide to a few of the mitigating measures

Protests at the Seattle WTO/OMC Ministerial Conference in 1999, in Prague (2000), at the 2001 G8 Summit in Genoa, in Cancún (2003) and in many other places and times less well publicised, have brought to a head the fundamental significance of trade, debt and development and the impact of economic globalisation. The significance of trade and debt to global development is certainly not new, and the staggering accumulation of Third World debt, and the central role of the industrialised nations in fuelling this crisis, from the 1960s are well-documented.[65] The movement of concern about and resistance to the effects of free trade (sometimes referred to as the 'movement of movements'[66]) is relatively recent, occurring in the same period as the rise to prominence of the anti-globalisation movement, and from a mixture of both sources a critique of tourism has grown.

There is little surprise that the myriad social movements protesting and advocating against globalisation have honed in on the impacts of trade liberalisation, unequal trading dependencies and the relationship between debt and poverty. The most recent manifestation of this movement is the 2005 'Make Poverty History' coalition and campaign which serves to emphasise the point that the debate is driven by poverty and its persistence. Not too surprisingly, the impacts of the prevailing economic model have prompted moves to consider fairer, more ethical, more responsible and more sustainable forms of trade and corporate governance[67] and alternatives to globalisation through the growth of the localisation movement.[68] These moves have included the tourism industry from which have sprung a range of initiatives designed to redirect the benefits of tourism and an increasing emphasis on tourism's potential benefits for the poor and the reduction of poverty. It is on this 'anti-poverty dividend' that any discussion of responsible tourism must focus in order to have meaning within the broader globalisation and development debates.

In the 1990s the business sector slowly awoke to the sustainable development agenda and to the role of responsible business in sustainability. A symptom of this shift has been the proliferation of codes of conduct. In recent years there has also been a steady growth in Corporate Social Responsibility (CSR) and triple bottom line reporting within which the impact of business in the economic, environmental and social sectors is accounted for and audited.[69] These new features of responsibility within industry have been experienced in the tourism industry as much as in others.

These moves have, however, been met with scepticism from both the business community and economists and observers critical of globalisation. The former chief economist at the OECD, David Henderson, asserts that 'widespread adoption of CSR would undermine the foundation of the market economy',[70] while commentators such as Naomi Klein[71] contend that to a large degree the acceptance of CSR has been designed to head off or respond to corporate public relations disasters or alternatively to retain market share and customer loyalty in the context of heightened consumer awareness. Of course, when applied to the tourism sector where profit margins tend to be exceedingly tight, there are also issues of how far CSR can reasonably extend beyond relatively glib codes of conduct and tourist education exercises.

As governments court globalisation they (and especially the powerful G8 governments) have tacitly endorsed the worst excesses of economic globalisation on the premise that they are powerless to do anything. And worryingly protestors against inaction, unethical and unfair trade policy have been dismissed as naive fractional anarchists. The absence of ethical international leadership has been astounding. To many critics of the prevailing structures of power, one thing would appear clear: left to its own devices the western industrial and financial edifice is incapable of genuine self-regulation for responsible behaviour, and the social and environmental consequences are ominous.

Codes of conduct, CSR and other techniques of regulating the industry in order to mitigate the ill-effects of neoliberal economic development are examined in this section in the context of the debate around imposed regulation against voluntary self-regulation. We start with a brief discussion of the general thrust of the tourism industry's attempts to control the debate around responsibility through the technique of voluntary self-regulation. We then briefly examine codes of conduct, corporate social responsibility and the recent development of fair trade in tourism. Only a few of these techniques are covered here as several, such as pro-poor tourism (covered in Chapter 3, 'Local Politics, Poverty and Tourism: The Micro Level of Responsibility'), and environmental auditing, ecolabelling and certification, holiday footprinting and carbon budget calculations are covered in Chapter 4 ('Tourism and the Environment: Eco by Name, Eco by Nature?'). All of these are seen, by the industry at least, as ensuring responsible actions on the part of industry players as well as being good public relations practice. On the other hand, critics tend to dismiss them as solely public relations exercises, or 'greenwash', offering only words as evidence of responsibility rather than actions.

Voluntary self-regulation

The issue of regulation of the tourism industry can be represented as a struggle for control of the industry between different interest groups. These may be many and varied as the industry itself is highly fragmented with many associations of hoteliers, travel agents, tour operators, caterers, transport companies and so on, at local, regional, national and international levels. Some of these groups undoubtedly have a role to play in promoting the attainment of ethical standards of practice, for the fragmented nature of the tourism industry is such that it would be impossible for all but the most bureaucratic of governments to regulate for all related practices and to enforce the legislation as well. As McKercher has pointed out, 'Effective control measures can only occur through integrated programmes that incorporate federal, state and local legislation'.[72]

Moreover, there is a clear and undisputed place for national and international legislation on safety matters relating to airlines and other aspects of tourism. In many other areas, however, regulation for sustainability and responsibility is a concept that is highly contested. This contest leads to the ongoing debate around the issue of self-regulation. It is essentially between two camps: those who believe that the industry should pursue voluntary self-regulation on issues relating to sustainability and responsible behaviour; and those who believe that regulation should take the form of statutory legislation imposed and enforced by government. In practice, it is not as simple as the descriptions of these two opposite camps would suggest – there are many combinations of company self-regulation, articles of association and government legislation which can be promoted and the issue may be more fairly and faithfully represented by a continuum of views between the two – but the argument is often presented as a simple dichotomy.

In the eyes of the industry and under the doctrine of the 'free market', constraints on the activities of different branches of the industry are for the worse. They inhibit competition and consequent price reductions, they create 'unnecessary' bureaucracy, they cause delays, they may alienate those who work in the industry, and they stifle its performance and effectiveness. But without these constraints, the industry is free to pursue profits with no regard to the external costs, the negative impacts on the environment, the culture, or the society. As Andrew Pendleton, head of trade policy at Christian Aid, says, 'allowing corporate social and environmental standard-setting to remain in the purview of voluntary initiatives leaves companies free to transgress with impunity – as many do'.[73] It is also free to use its voluntary attempts at self-regulation as a public relations exercise or marketing ploy. And without official, non-industry-based monitoring and inspection, it can deceive its consumers into believing that its operations are environmentally friendly, ethically sound or responsible.

At the same time, the fragmentation of the industry is one of the factors which permits operators within it a certain inconsistency. When government tries to regulate the operations of private companies, there are few industries in which it is allowed to do so without vociferous opposition from companies which vigorously uphold the benefits of voluntary self-regulation. Representing business, the Confederation of British Industry (CBI), for example, has consistently lobbied government and international institutions against the approval of rules which would make UK and other companies improve their social and environmental performance. When tourism companies are asked about the industry's responsibilities, on the other hand, their answers are somewhat at variance with their attitudes to regulation. As a Tourism Concern/WWF study of a number of tourism companies showed:

'All operators stated that national governments had some responsibility, and nearly 60 per cent of operators said that governments had total responsibility. This view was echoed by travel agents, carriers and hotels'[74] and clearly points to the need for some form of authorised regulation rather than voluntary self-regulation.

The Green Globe scheme, launched in 1994 by the World Travel and Tourism Council (WTTC), is an international environmental management and awareness programme designed to encourage travel and tourism companies, whatever their size, sector or location, to make a commitment to continuous environmental improvement. It offers achievement awards, training, networking, publications, advisory and information services, branding and marketing assistance, and was expanded in 1999 to include independent auditing and again in 2001 to include actual measurement of environmental improvements through benchmarking on an annual basis. The scheme has numerous participating companies and organisations in the LAC countries such as the Sandals Resorts chain, the Italian Venta Club chain of all-inclusives and the Caribbean Alliance for Sustainable Tourism. The establishment of Green Globe by the WTTC was seen by the proponents of self-regulation as helpful because 'it enabled tourism companies to seek advice on environmental matters from experts who also understood the travel trade . . . Furthermore, such progress had been achieved, by and large, without the inter-vention of outside bodies. This demonstrated the commitment of the tourism industry to sustainable tourism.'[75]

Brian Wheeller[76] is not the only one to argue, however, that the WTTC represents business interests which advocate the message 'no outside regulation, we can regulate ourselves', and who are acting, not altruistically, but only in the immediate interests of their members. The point about Green Globe is that it is there 'to prevent, by having in place a self-regulatory system, any government interference in the workings of the industry'.[77] Corporate players in the tourism industry are aware that increased public and media attention to their means of operation will interfere with the market dominance that they currently enjoy. With the Green Globe scheme, as Wheeller says, tourism corporations are 'able to hide behind the façade of self-regulation'.[78] To obviate such criticism, Green Globe 21 (as it is now known) established Green Globe Accreditation (GGA) which it describes as 'an independent organisation that has been given the contractual responsibility, . . . for assessing and accrediting organisations seeking to offer Green Globe 21 assessment services'. GGA's supposed independence only means that it was established as a separate entity from Green Globe 21 itself. It does not represent independence and it does not offer any form of governmental oversight of operations. Indeed, it was set up precisely to avoid interference from government.

It is perhaps too easy to interpret the WTTC's wholesale encouragement of self-regulation as promoted cynically, purely in self-interest and in pursuit of short-term profits. It could also be interpreted as a genuine attempt to help the industry adapt to what may become in the future environmentally and socially essential regulation. The former interpretation suggests that business is simply trying to avoid the inevitable; the second that it is prescient in trying to adapt to it. Whichever interpretation is taken, Haslam believes that the real world imperfection of the

Latin American context 'increases the need for legal (and moral) obligations not decreases it'.[79]

Whether government legislation would really help to reduce the uneven and unequal nature of tourism development may be debatable. But voluntary self-regulation led by bodies such as the WTTC and the WTO/OMT, whose stated aims are the promotion of the tourism industry rather than its restraint, is likely to lead to policies which further the pursuit of profits in a business world where profit maximisation and capital accumulation form the logic of economic organisation. Government legislation is likely to curb the tendency of such policies to compromise human rights and labour and environmental standards.

Codes of conduct

In the 1990s codes of conduct became an important feature of the tourism industry. They attempted to influence attitudes and modify behaviour and embodied the industry's efforts to make itself more responsible. Almost all codes are voluntary; statutory codes, backed by law, are very rare. But to some extent, and especially at the more global scale, codes of conduct have been overtaken by the more recent and fashionable ecolabels, certification schemes, benchmarking and corporate social responsibility schemes. Codes of conduct are still important, however, at the scale of the local and single utility, and in special circumstances where, for instance, the protection of specific ecosystems is deemed necessary.

Despite the general criticism of blandness, some codes of conduct are very impressive in their range of issues and in their depth of discussion and information. There exist, however, a number of problems associated with the use of codes of conduct: their monitoring and evaluation; the conflict between codes as a form of marketing and codes as genuine attempts to improve the practice of tourism; the debate between regulation or voluntary self-regulation of the industry; and the variability between codes and the need for co-ordination.

The World Travel and Tourism Environment Research Centre (WTTERC) established a database of codes of conduct relevant to the travel and tourism industry, and was prompted to do so by their proliferation, especially in the international sphere. Despite this database, in 1995 research suggested that most codes offered no measurable criteria and conformed to no widely accepted set of standards.[80] The study further suggested that many codes were devised by tour operators simply to offer a palliative to the pressure exerted by the environmental lobby and in anticipation of potential criticism. In 2001 Naomi Klein described the 'proliferation of voluntary codes of conduct and ethical business initiatives [as] a haphazard and piecemeal mess of crisis management'.[81] She concluded that codes as a form of self-regulation gave to those who devised them 'unprecedented power . . . the power to draft their own privatised legal systems, to investigate and police themselves, as quasi nation states'.[82] Klein was referring especially to TNCs, but the criticism might equally apply to many small-scale tourism businesses which devise their own codes of conduct. Many codes self-designed by the user, few if any of which are monitored by independent bodies, are the epitome of self-regulation, and are at best questionable and at worst meaningless – tantamount to asking the metaphorical bull to tread carefully in the china shop. So what meaning and purpose can we find for these codes?

Codes of conduct may serve as attractors to potentially discerning and critical customers who are seeking a relatively ethical holiday, allowing them still to travel to faraway and exotic places, have a holiday and return environmentally enriched. In these senses, the code of conduct becomes more a marketing ploy than a set of standards by which to guide a company's behaviour and practices. The marketing ploy works as 'spin' to show an environmentally responsible attitude to the supposedly discerning consumer or to head off potential antagonism and bad publicity from critical environmentalists. In both cases this belies their supposed true purpose of practising tourism responsibly. Naomi Klein describes corporate codes of conduct as the 'most controversial by-product of brand-based activism',[83] drafted as they are by public relations outfits wooing consumers or steadying the corporate ship in the aftermath of public scandals.

In March 2000, the WTO/OMT, along with UNEP and UNESCO, launched the Tour Operators' Initiative (TOI) for Sustainable Tourism Development,[84] to which many UK tour operators have committed themselves. The initiative is designed to promote protection of the environment (in its widest sense), co-operation with local communities, respect for local cultures and compliance with local laws and customs. The Initiative includes a collection of Best Practice case studies, one of which is that of the company TUI Nederland, one of the world's biggest tourist corporations and, as it describes itself 'the market leader in the Dutch travel industry'. The specific example cited by the TOI is TUI's scheme for 'Promoting Responsible Travel in Curaçao and Bonaire',[85] one element of which is a series of diving guidelines formulated for the project. These are given in Box 2.3 and have been adopted by almost all dive operators on the two islands. There is no intention here to deny the claims made in the TOI or by TUI Nederland and still less to decry what is obviously vital work on the two islands of Curaçao and Bonaire, but it is interesting to note that in December 2003 TUI's director of environmental management, Wolf Michael Iwand, gave rise to a certain degree of scepticism among other delegates at a conference on 'Tourism: Unfair Practices – Equitable Options'. In response to the question 'why should the tourism industry have an interest in poverty alleviation?', Iwand stated that 'the business of business is business', betraying that there is no such interest in poverty within the business sector and that TUI do not consider poverty alleviation as a part of their operational objectives. On the contrary, regarding inequality he stated that 'we are living comfortably with it. People in the destinations are asking: send us more [tourists]'.[86] Indeed, people in the destinations do cry out for more tourists, but usually because they have already been sucked into tourism as their only means of making a living.

Another of the best practice case studies is TUI Nordic's implementation of the code of conduct devised by ECPAT (End Child Prostitution and Trafficking) to combat the sexual exploitation of children within the tourism industry, detail and discussion of which are given in Chapter 7 and Box 7.6.

Box 2.3   TUI's Diving Guidelines for Bonaire and Curaçao

1   Guests must receive an orientation and explanation of eco-diving standards, including:
   - Maintain good buoyancy control;
   - Practice good finning technique and body control;
   - Ensure all equipment is adequately secured;
   - Avoid all contact with the bottom and/or marine life;
   - Handling, manipulation and feeding of marine life to be conducted under expert guidance only;
   - Take nothing except recent garbage out of the water; and
   - Restrict use of gloves and kneepads whilst reef diving.
2   Dive operation staff must practice eco-diving standards.
3   Corals and other marine life may not be offered for sale.
4   Environmental materials (e.g., marine life ID) must be available to guests and staff.
5   The dive operation must promote a strict 'no contact' policy for reef diving.
6   Pre-dive briefings must be given, reminding guests of eco-diving standards.
7   Dive boats must use moorings, drift dive or hand place anchors in reef environments.
8   Dive boats must have adequate garbage facilities aboard.
9   Dive boats must have a 'minimal discharge' policy in force.
10 Dive operators on Curaçao must be members of the Curaçao Dive Operator Association (CDOA) and actively support the CDOA voluntary dive-tag programme benefiting marine life management and conservation projects. Operators on Bonaire must actively support the Bonaire National Marine Park, in addition to abiding by all local environmental laws, regulations and customs.

Source: www.toinitiative.org/good_practices/case_studies.htm.

Examples of projects carried out by the member companies of the TOI (such as the Scandinavian Leisure Group, Japan Travel Bureau, Finnair Travel Services, Orizzonti and Premier Tours) are impressive. But one of the problems of best practice examples is the standard output of such initiatives. They also fail to give a balanced or critical analysis of the general practice of the industry. And this type of output can still be used as a means of marketing, lending false moral and ethical high ground to those whose principal and overriding aim is to make a financial profit.

Journey Latin America (JLA) is a British tour operator specialising, as its name suggests, in Latin American destinations for its clients. On its website JLA promotes a Traveller's Code[87] which covers issues such as photography, souvenirs, haggling, litter, conservation of resources and learning about the holiday destination before departure, amongst other things. It would be fair to say that this figures prominently in its website, and it might be reasonably presumed that the code is distributed to all its clients when they make a booking. In fact this is not the case. Furthermore, when the company was telephoned[88] for the purposes of general enquiries about JLA, it quickly became clear that the first two employees contacted (one in flight bookings and one in tours) were not even aware of the existence of the code. Again, this suggests the possibility that the code is used purely as a palliative or as a set of false environmental credentials.

As a separate initiative, endorsed by the United Nations General Assembly in 2001, the WTO/OMT launched the Global Code of Ethics for Tourism which sets out ten points based on principles of sustainability with special emphasis on involving local communities in planning, managing and monitoring tourism developments. It includes nine articles outlining the 'rules of the game' for destinations, governments, tour operators, developers, travel agents, workers and travellers themselves. The tenth article covers implementation of the code.[89]

Initiatives such as the TOI and the Global Code of Ethics for Tourism may help to overcome some of the problems associated with codes of conduct and to inject a little rigour into their application; and the Best Practice case studies clearly show that for specific purposes in specific locations or circumstances their use may be both appropriate and effective. Despite the fact that such initiatives have done little to reduce the industry's tendency to use codes as a means of marketing to lend false moral and ethical claims to advertising in order to attract more custom, for the sake of the intending tourist, it would certainly seem worthwhile to consult the appropriate website before travelling – see Appendix.

A suitable system of reference for codes of conduct, made easily accessible to tourists and others who wish to verify the claims made by any member of the industry, would help to improve the credibility that can be placed in codes of conduct. But any such system would need to take account of who initiates them, for whom they are intended, and who monitors their use if indeed any assessment is made of their effectiveness, which it rarely is. To this end some system of environmental auditing of codes of conduct would be helpful. The WTO/OMT or WTTC are best placed and best resourced to achieve this, but these industry-based organisations represent the interests of their members and companies which provide their funding.

Until such time as an effective co-ordination and evaluation system for codes of conduct is devised and operated, it seems likely that most individual tourism companies will continue to produce their own sets of guidelines. Examples of these can be found for many companies on their websites. For the sake of illustration, we briefly discuss two UK-based companies, Explore Worldwide and Exodus, both of which run tours to various destinations in the LAC countries, and whose respective websites are given in this chapter's notes.[90] Under the heading 'Responsible

Tourism', both companies outline policies on litter and waste disposal, wildlife protection, local sourcing of supplies and personnel, respect for local culture and other features of their activities. The Exodus guidelines are particularly detailed, and for illustration the first part of the company's Responsible Tourism Policy statement is given in Box 2.4. The statement goes on to give many more guidelines for their clients, the tourists. For both companies, however, apart from soliciting feedback from their clients, their documentation and websites publicise no formalised evaluation and monitoring of their codes; and although both companies display their awards and affiliations – in November 2004 Exodus won the overall award for Responsible Tourism at the travel industry's World Travel Market – no yardstick by which the effectiveness of their codes can be judged is given.

Finally on codes of conduct in the tourism industry, it is worth noting that a Tearfund 2001 report[91] suggested that many companies in the survey conducted for their report produced a responsible tourism policy which was effectively just a sheet of suggestions on how tourists should behave. Such efforts serve to shift responsibility away from the company on to the tourists.

Corporate Social Responsibility (CSR)

Corporate Social Responsibility (CSR) is a technique which is promoted by relevant supranational organisations such as the WTO/OMC, WTO/OMT and many of the United Nations agencies. It has been adopted by many companies and promoted by NGOs such as Tearfund and Fair Trade in Tourism, an offshoot of the UK's Tourism Concern. The International Business Leader's Forum (IBLF) has also established a CSR Forum to promote the technique among its members. As the director of South American Experience has stated, CSR is currently very fashionable: 'We had these same responsible policies years ago, but now we use the label "corporate social responsibility"'.[92]

The Tearfund, one of the UK's leading relief and development agencies, explains CSR in the following terms: 'Once there was just the financial bottom line. Now, companies recognise they must be accountable for the way they affect people, the community and the environment – the new triple bottom line.'[93] It further suggests that the trend towards CSR is inexorable, although the New Economics Foundation reports that, while a third of FTSE[94] companies issued environmental reports, in 2000 only 4 per cent produced fully verified reports.[95]

As the term suggests, Corporate Social Responsibility is very much intended for corporations, those companies large enough and wealthy enough to fund departments which monitor and report on their environmental, social and community impacts. Despite this, the technique is strongly promoted by many NGOs, among them Fair Trade in Tourism which suggests that the technique of CSR emerged in the late 1990s out of NGO efforts to create 'a more equitable international trade system'.[96] Haslam suggests that in the LAC countries, 'The importance of NGOs in advancing the CSR agenda is surprisingly strong'.[97]

The possibility of changing management practice through CSR exists, and Fair Trade in Tourism suggests that: 'Experience in a range of business sectors shows

Box 2.4   Extracts from Exodus's Responsible Tourism Policy

We will use locally owned and run services by preference.

- We use locally owned and run accommodation by preference.
- We will aim to where possible purchase our equipment and food from locally produced source.
- We support local income generation and small business enterprises by supporting locally owned shops and restaurants, and using local guide services.

We encourage clients through our leaders, the literature and presentations to act in a responsible way and with an insight and understanding of the host destination.

We will work together with our business partners in the host destinations to implement the highest of standards and where necessary work out an agenda on how these standards can be met.

We will take into consideration economic, environmental and cultural issues when looking at new destinations.

We will aim to ensure that our type and scale of tourism is appropriate to local conditions and operate within the limits set by local appropriate infrastructure and carrying capacity.

We aim to take responsibility for maintaining and improving the environment

- We aim to minimise water and atmospheric pollution from tourism developments.
- We aim to achieve zero litter policies.
- We aim to integrate environmental considerations into all economic considerations.

We will provide regular and ongoing training, in the principles and practices of responsible tourism, both to our staff in the office and those employed out of the office.

We will try to ensure that foreign operators and guides are not used in preference to local operators and guides.

We will provide financial support for local economic and environmental projects that will benefit the local communities, thereby actively encouraging local community involvement in Tourism Projects.

UK operations
Our operations within the UK are conducted within the same framework of responsibility that guides our overseas operations.

> We recycle 90% of office waste paper and recycle glass and cans . . .
> Through tight management systems, we try to reduce paper consumption and
> keep waste to a minimum. We also recycle ink and toner cartridges . . .
>
> We monitor our energy consumption, in the office, against the Carbon
> Trust's benchmarking system and score well against the average Kwh per
> head . . . We buy our electricity through companies which are offering the
> highest percentage of energy generated from renewable sources.
>
> *Source*: www.exodus.co.uk/restourism.html, accessed 25 August 2006.

that CSR and ethical trade practice can be put into action without jeopardising
profit levels and share prices; indeed, they can actually enhance these. The tourism
industry has an excellent opportunity to take this on board.'[98] We have already
noted the possibility that exists for companies to promote such actions as CSR as
public relations activities designed to enhance their image without necessitating a
fundamental change in practice – a practice commonly referred to as 'greenwash'.
To adopt such a perspective on all such efforts without investigative research would
clearly be overly cynical and we would not recommend such an approach here.
Enough cases of 'greenwash', however, have been highlighted to temper vigorous
enthusiasm for CSR and similar techniques. Haslam has pointed out that many
NGOs in the LAC countries which promote CSR receive funds from organisations
such as the Inter-American Development Bank, the Kellogg Foundation, the Inter-
American Foundation and national development agencies, prompting the question
'whose agenda is the CSR agenda? Is it homegrown or imported from abroad? . . .
If Latin American NGOs know that there is money for CSR, they will create
programmes to access it.'[99]

In January 2001 Tearfund published a report into the responsible business
practices of 65 UK-based tour operators, and found that 'the tourism industry has
made some good progress on environmental issues but lags behind other industries
in terms of fulfilling its social and economic obligations'.[100] It also found that 'only
half of the companies questioned have responsible tourism policies, and many of
these are so brief as to be virtually meaningless'.

The Tearfund also concludes that currently it is the smaller operators which offer
a more ethical experience to holidaymakers, 'paying a higher proportion of profits
to charity, offering more training to local operators and developing more local
partnerships'.[101] In 2001 the Association of Independent Tour Operators (AITO),
which deals with the smaller and more specialised tour operators, developed
Responsible Tourism Guidelines which cover general good practice in sales,
marketing and pre-departure information. The AITO chose to develop its own
guidelines for its member companies after its decision that over the years the Green

Globe initiative (to which it had been affiliated) had become too corporate in style for the AITO.[102]

The AITO's implicit criticism of the corporate agenda associated with the Green Globe initiative is repeated elsewhere in its website where it claims that its own members are 'all independent of the vertically integrated groups within the industry'.[103] This distinction is important to the AITO because it allows its members to distance themselves from the widely perceived unsustainability associated with the practices of the large-scale, mainstream, corporate tour operators. In turn, this distance not only gives them a status of specialists offering personal service but also allows them to claim, almost by default, environmental and ethical principles for their own. There is no doubt that AITO values its association with responsible tourism practices highly and it promotes a code of conduct for tourists along with the Campaign for Real Travel Agents (CARTA). It is, however, an organisation designed to help its members market themselves and improve their performance (maximise their profits). Whether this aspect of its work conflicts with its promotion of responsibility and sustainability is an open question. One of its guidelines for tourists, for example, is 'Get your holiday off to a green start – if possible, travel to and from your airport by public transport.'[104] Despite the fact that the most polluting element of the holiday is likely to be the air journey, the code of conduct ignores this leg of the journey because it is essential to the holidays that it is selling, and instead the AITO chooses to focus attention on a relatively trivial leg of the journey.

But the growing acceptance of CSR by the industry and its promotion by government departments and non-governmental organisations indicates a degree of change within the industry, perhaps not to the extent of a shift in major purpose, but a change nevertheless. The IBLF also suggests that the industry is changing: 'In the mid-1990s, for example, its focus on issues of business ethics, social development and human rights was seen as rather radical and ill-focussed. Less than six years later, the same issues could be discussed comfortably in most board rooms.'[105] Whether such discussions will bring about the change desired by so many NGOs and communities and by some relatively forward-thinking company executive boards is much more questionable.

Fair trade in tourism

Fair trade has worked most effectively to date with simple and tangible commodities such as coffee and chocolate. Because wages and the share of total profit are usually lowest at the source of production (the small growers, pickers and packers), fair trade seeks to achieve a realistic wage per production cost for the most economically marginalised and vulnerable in the production and logistic chain. It is a challenge to traditional economic theory and practice in that it seeks to set a price based on principles other than those of pure profit maximisation and on practices other than seeking the lowest cost of production in so-called efficient markets.

In the tourism industry there is a number of initiatives that have some resonance of 'fairly traded tourism' particularly through the 'awareness holidays' run directly or promoted by a growing number of non-governmental development organisations such as Oxfam and ActionAid. And there are elements of tourism-related activities, such as the production and sale of local handicrafts that provide the opportunity for elements of fair trade.[106] But the neoliberal economic model is not designed to operate on principles other than profit maximisation, and one of the institutions most strongly embodying this principle, the World Trade Organisation, is well known for 'its incapacity to be inventive when it comes to trying to marry sustainable production, fair trade and ethical consumption with international trade policy'.[107] The current rules of international trade make it impossible to exercise any kind of positive discrimination in favour of more responsible and sustainable trading links.

In one of the few attempts to provide a consolidated understanding of fair trade in tourism, Cleverdon and Kalisch observe: 'Little or nothing is known about fair trade in services, let alone fair trade in the hospitality sector.'[108] Involving primary research and including a dialogue with tourism providers from Third World countries, their work provides one of the few focused insights.

Cleverdon and Kalisch document a number of key differences between fair trade in services and primary commodities as a means of identifying the potential and shape of fair trade in tourism. Perhaps the most significant difference is that fair trading organisations are non-profit making, whereas those small independent tour operators who are practising some quasi-fair trade elements (such as direct trading relationships) also need to ensure commercial viability. Price is a significant element here. Whereas the tourism industry depends on low prices that are flexible according to both time and place, the price of products like coffee can be gauged in relation to a fixed world price. This begs the questions as to whether price should be the main determinant of fair trade tourism and whether there are other attributes such as the degree of local control or the distribution of benefits that may be more significant. There are also issues concerning the lack of collective organisation among Third World tourism producer communities, contrasted to the degree of collective organisation exhibited by primary commodity producers.

As prerequisites for fair trade in tourism, Cleverdon and Kalisch identify a number of variables. These are certainly not unique to the tourism sector, and must be understood within the context of a broader discussion of development. As such they are reflective of the causes of unevenness and inequality that condition the activity of tourism. The prerequisites are 'access to capital, ownership of resources, distribution of benefits and control over representation of the destination in tourist-generating countries, and it needs to ensure transparency of tourism operations, including price and working conditions'.[109] In short, this brew of fair trade in attempting to 'provide a better deal for producers or, in the case of tourism, service providers in the South' has reflections in new forms of tourism and is not dissimilar to pro-poor tourism initiatives,[110] which we cover in Chapter 3.

In the context of economic globalisation and growing, not diminishing, inequalities, commentators have been quick to acknowledge the weaknesses and shortcomings of the fair trade movement. Small changes can undeniably make a significant positive impact to the lives of those involved. But the degree of change is likely to be minuscule in its impact on trading patterns and trends, at least for the foreseeable future. For example, pro-poor tourism is in its infancy, and in terms of national poverty reduction it has had little impact,[111] and clearly tourism cannot be regarded as a panacea for rural development. (There is little literature on pro-poor urban tourism though it has begun to emerge in European cities involving immigrant communities – see Chapter 6.) As Cleverdon and Kalisch observe, fair trade products constitute around 3 per cent of the commodity market, and there are no reasons to believe that fair trade tourism would comprise anything but a small percentage of the total global tourism industry; though the percentage contribution in individual countries can vary considerably.

This raises questions as to the relative merits of fair trade (small-scale exemplary holidays, for example) as opposed to CSR and ethical trade in tourism (involving corporate codes of conduct, regulation of the industry and compliance with minimum standards), or ultimately to localisation and a reduction in western consumption (essentially travelling less). Although Hutnyk appears to sympathise with the good intentions of fair and alternative trade advocates, he concludes: ' "Alternative" travel, just as much as the alternative trade promoted by many organized aid groups, works as a reassuring front for continued extension of the logistics of the commodity system, even as it masquerades as a (liberal) project of cultural concern.'[112] And from an ethical perspective there are question marks against the promotion of elite 'ethical' niche markets, consumption patterns and trading systems that may reinforce, rather than challenge, global inequality.

So is responsible tourism possible?

The structure of power underlying the tourism industry, how it developed historically into its current shape and how this structure influences the activities and operations of the industry provide the key to understanding the prospects and limitations of responsible tourism and its ultimate contribution to development.

The prevailing form of economic development pursued and promoted so vigorously by the Washington Consensus and its supranational agencies along with the G8 governments, namely globalising free trade, is now widely but perhaps unwisely accepted as 'the only game in town'. The currently emerging challenge to this consensus and acceptance of it – the challenge represented especially by Presidents Hugo Chavez and Evo Morales of Venezuela and Bolivia respectively – has so far managed little in the way of change to the prevailing economic and trade practices within the LAC countries apart from Venezuela and Bolivia; and in the case of the latter there is still much doubt about the degree of change that the country may undergo as a result of the challenge.

Through a variety of mechanisms and policies, including debt restructuring, privatisation and deregulation, the prevailing form of economic development asserts the primacy of economic growth, capital accumulation and investment and offers the model of the wealthy technocratic nations for all others to follow. At the same time, however, it locks Third World nations into a dependency upon First World finance, First World technology, First World communications, First World infrastructure, transnational corporations based in the First World and, still, largely First World visitors. This dependency, even stranglehold, obliges Third World nations to sell their assets – natural resources – in order to gain foreign earnings that they can use to pay off their debt and to enable them to qualify for further loans in order to develop in a First World style that leaves them in even greater debt. A vicious circle. These assets include their pristine ecosystems, if they have any left, their tropical heat, their beaches and their distinct cultural characteristics, all of which can be exploited for the dollars to be earned from tourism, which acts for national governments of LAC countries as the newest addition to the industries that they can develop in order to gain foreign currency. The new forms of tourism, such as ecotourism, responsible and ethical tourism, are also promoted by the supranational agencies and require the same transnational investment as other more traditional industries.

So, is responsible tourism possible? The new forms of tourism to which we refer are, by and large, associated with the notions of responsibility, sustainability and ethical behaviour. But it is clear that the prevailing economic model of development causes and creates considerable problems and challenges in the LAC countries, problems associated particularly with poverty and inequality and in such challenging circumstances it is rarely easy to promote responsibility. In Third World environments, industries which operate in the globalised system and which follow its norms have a tendency to be exploitative, whether it be of natural resources or of human populations; and tourism is no exception in this regard. Countless instances of the exploitative nature of tourism developments in the Third World have been documented over the last two decades. The new forms of tourism, however, are intended to overcome such exploitation. As such they have aligned themselves with many of the techniques and policies used to mitigate the ill-effects of globalising economic growth – codes of conduct, corporate social responsibility, fair trade in tourism, and others which we cover in later chapters. Many of these techniques are used to mitigate the problems caused by the prevailing model of development and are seen by some as a means of obviating the need for regulation of the industry. On the other hand, the CORE Coalition of NGOs (which includes Friends of the Earth, Tourism Concern, Christian Aid, Amnesty International and the World Wide Fund for Nature) has consistently lobbied government to make companies legally responsible for corporate social responsibility. A recent UK government White Paper,[113] however, argues for allowing companies to take a voluntary approach to CSR. The outcome of the debate will reflect the balance of power between the different groups, civil society (represented by the NGOs) and business society (represented by the companies); but a joint UK and Nicaraguan research team note that 'years of lobbying has not changed the structure of the world economy nor the ideology of its ruling institutions on the alternative views of development advocated by NGOs'.[114] Such a pessimistic interpretation inevitably comes from a group of academics. Activists, on the other hand, would be quick to point out a host of examples to show that lobbying and constant pressure can bring positive results.

Notes

1 Weinberg, B. (1991) *War on the Land: Ecology and Politics in Central America*, London: Zed Books, p. 6.
2 Weinberg, B. (1991), p. 6.
3 Galeano, E. (1973) *Open Veins of Latin America*, London: Monthly Review Press, p. 72.
4 It should be noted, however, that the exactness of these statistics is contested by various scholars.
5 Offspring of Spanish and American Indian.
6 Offspring of a white person and a negro.
7 Now published by Footprint Books, Bath, UK.
8 Britton, S. (1982) 'The Political Economy of Tourism in the Third World', *Annals of Tourism Research* 9: 331–8.
9 William Blum (2003) *Killing Hope: US Military and CIA Interventions since World War II*, London: Zed Books, p. 229.
10 Both mass enclave tourism and cruise ship tourism are covered only in passing in this book. We therefore recommend the reader to Polly Pattullo's second edition of *Last Resorts: The Cost of Tourism in the Caribbean*, published by the Latin America Bureau and Cassell in 2005.
11 John Vidal (2004) 'Eco-holidays Fail to Take Off', London: *The Guardian Travel*, p. 14. 1 May.
12 Owing to possible confusion with the initials WTO, signifying both the World Trade Organisation and the World Tourism Organisation, when both organisations are discussed together it is common practice to link the English initials with the Spanish and French initials. These are respectively OMC for the World Trade Organisation and OMT for the World Tourism Organisation. Thus the World Trade Organisation is abbreviated to WTO/OMC and the World Tourism Organisation to WTO/OMT. We follow this practice.
13 Severino Resende, quoted in 'Our Agenda for UNCED', *UNEP-UK News*, Spring 1992, p.8.
14 Krishna B. Ghimire (2001) *The Native Tourist: Mass Tourism Within Developing Countries*, London: Earthscan.
15 Hettne, B. (1995) *Development Theory and the Three Worlds: Towards an International Political Economy of Development*, Harlow: Longman, p. 25.
16 Sachs, W. (1999) *Planet Dialectics: Explorations in Environment and Development*, London: Zed Books.
17 After Hernando de Soto (2001) *The Mystery of Capital: Why Capitalism Triumphs in the West and Fails Everywhere Else*, London: Black Swan, p. 241.
18 Rist, G. (1997) *History of Development*, London: Zed Books, p. 2.
19 Department For International Development (2000a) *Eliminating World Poverty: Making Globalisation Work for the Poor*, London: The Stationery Office, p. 23.
20 Department For International Development (2000b) *Halving World Poverty by 2015: Economic Growth, Equity and Security*, London: Department For International Development, p. 9.
21 Yao Graham (2005) '. . . Only If We Stay in the Market', London: *The Guardian*, 12 March 2005.
22 Rist, G. (1997), p. 239.
23 Arturo Escobar (1995) *Encountering Development: The Making and Unmaking of the Third World*, Princeton: Princeton University Press, p. 4.
24 Eric Hershberg (2003) 'Latin America at the Crossroads', in NACLA Report on the Americas, XXXVII, 3, November/December, New York: North American Congress on Latin America, p. 20.
25 James Petras (2006) 'Myths and Realities: Is Latin America Really Turning Left?'

CounterPunch, 3 June 2006, www.counterpunch.org/petras06032006.html, accessed 22 August 2006.

26 Kofi Annan (2003) Opinion Column, London: *Guardian Special on Trade*, 8 October, p. 9.

27 *Ibid.*

28 Gordon Brown and Jim Wolfensohn (2004) 'A New Deal for the World's Poor', London: *The Guardian*, 14 February.

29 Gamini Corea, former Secretary General of UNCTAD, in *UNEP-UK News*, Spring 1992.

30 The G8 countries are: Canada, France, Germany, Italy, Japan, Russia, the UK and the USA.

31 George Gelber quoted in Charlotte Denny (2004) 'Brown Confronts Aid Target Critics', London: *The Guardian*, 16 February.

32 See for example: War on Want (2005) 'UK Government "Part of the Problem" for Developing Countries in 2005', www.waronwant.org; and World Development Movement, War on Want, Friends of the Earth, and National Union of Students Report (2005) '2005 and Sustainable Development: Why the UK Government is Part of the Problem'.

33 Agencia Informativa Latinoamericana Prensa Latina (2006), www.plenglish.com.

34 Rostow, W. (1960) *The Stages of Economic Growth: A Non-communist Manifesto*, Cambridge: Cambridge University Press.

35 Butler, R. (1980) 'The Concept of a Tourist Area Cycle of Evolution: Implications for Management of Resources', *Canadian Geographer* 24, 11: 5–12.

36 Rist, G. (1997).

37 WTO/OMT (1995) *GATS and Tourism*, Madrid: WTO/OMT, p. 1.

38 Arthur MacEwan (2001) 'The Neoliberal Disorder: The Inconsistencies of Trade Policy', *NACLA Report on the Americas* 35, 3: 28. New York: NACLA.

39 Bridget Wooding and Richard Moseley-Williams (2004) 'Worlds Apart', *Interact*, Spring. London: Catholic Institute for International Relations, p. 9.

40 Fred Rosen (2003) 'Changing the Terms of the Debate: A Report from Antigua', *NACLA Report on the Americas* XXXVII (3), November/December, p. 25. New York: North American Congress on Latin America.

41 Wayne Ellwood (2004) 'The World Trading System', *New Internationalist* 374, December, Oxford: New Internationalist.

42 Bernard Crick (2004) 'How the Rich Stole the Dream', London: *The Independent*, 11 June. Crick was referring to the manipulation of the capitalist tax system to benefit the rich, but in the context of the development of the capitalist economic system. We consider that the contexts of his remark and of ours are similar enough for it to be used to describe deregulation.

43 Martin Khor (1996) 'Global Economy and the Third World', in Jerry Mander and Edward Goldsmith (eds) *The Case Against the Global Economy*, San Francisco: Sierra Club Books, pp. 47–8.

44 With thanks to Paul Gonsalves (1995) 'Structural Adjustment and the Political Economy of the Third World', *Contours* 7, 1: 33–9, from whose example our own example was adapted.

45 David Korten (1996) *When Corporations Rule the World*, West Hartford, CT: Kumarian Press, p. 54.

46 Deborah McLaren (2003) *Rethinking Tourism and Ecotravel: The Paving of Paradise and What You Can Do to Stop It*, Bloomfield, CT: Kumarian Press.

47 Mariama Williams (2003) 'The Political Economy of Tourism Liberalisation, Gender and the GATS', www.genderandtrade.net/GATS/.

48 World Tourism Organisation website (2004) 'Liberalisation with a Human Face', www.world-tourism.org/liberalization/trade_services.htm.

49   *Ibid.*
50   Williams (2003).
51   Larry Elliott (2004) 'Aid Flows Are a Bridge to Fairer World Trade', London: *The Guardian*, 31 May, p. 23.
52   Central Intelligence Agency (2000) 'Global Trends 2015', United States Government. Quoted in 'States of Unrest III', a report by the World Development Movement, London, 2003, p. 5.
53   Vandana Shiva (1999) 'The Round to the Citizens', London: *The Guardian*, 8 December, p. 8.
54   Joe Bandy (1996) 'Managing the Other of Nature: Sustainability, Spectacle and Global Regimes of Capital in Ecotourism', *Public Culture* 8, 3, Spring.
55   World Development Movement (2002) *GATS: A Disservice to the Poor*, London: World Development Movement, p. 5. Among other things, the WDM cites a 1997 speech made by the WTO/OMC's Director of Trade in Services Division at a conference entitled *Opening Markets for Banking Worldwide: The WTO General Agreement on Trade in Services*.
56   European Community (2000) 'Opening World Markets for Services: Towards GATS 2000', http://gats-info.eu.
57   The G7 nations are the G8 (see note 30) minus the Russian Federation.
58   WDM (2002), p. 54.
59   Angela Kalisch (2001) *Tourism as Fair Trade: NGO Perspectives*, London: Tourism Concern, p. 4.
60   Victor Menotti, Director of the Environment Programme of the International Forum on Globalisation: letter to Equations, Indian NGO campaigning on tourism issues, June 2003.
61   Kalisch (2001), p. 5.
62   *Ibid.*, pp. 5–6.
63   K. T. Suresh (2004) 'Why the GATS Is a Wrong Framework through which to Liberalise Tourism', Equations Policy Briefing at the WTO/OMT International Symposium on Trade in Tourism Services, Madrid, March.
64   Barry Coates (2001) 'GATS', in B. Bircham and J. Charlton (eds) *Anti-Capitalism: A Guide to the Movement*, London: Bookmarks Publications, p. 29.
65   See for example Susan George (1988) *A Fate Worse than Debt*, London: Penguin, and Payer, C. (1991) *Lent and Lost: Foreign Credit and Third World Development*, London: Zed Books.
66   Notes From Nowhere Collective (2003) *We Are Everywhere: The Irresistible Rise of Global Anticapitalism*, London: Verso.
67   Starkey, R. and Welford, R. (eds) (2001) *The Earthscan Reader in Business and Sustainable Development*, London: Earthscan.
68   Colin Hines (2000) *Localisation: A Global Manifesto*, London: Earthscan.
69   Elkington, J. (2001) 'The "Triple Bottom Line for 21st-century Business"', in R. Starkey and R. Welford (eds) *The Earthscan Reader in Business and Sustainable Development*, London: Earthscan.
70   Charlotte Denny (2001) 'Profit Motive', London: *Guardian Weekly*, 15–21 November, p. 27.
71   Naomi Klein (2001) *No Logo*, London: Flamingo.
72   McKercher, B. (1993) 'Some Fundamental Truths about Tourism: Understanding Tourism's Social and Environmental Impacts', *Journal of Sustainable Tourism* 1, 1: 11.
73   Andrew Pendleton (2004) 'Beyond the Bottom Line', London: *The Guardian*, 12 June.
74   Forsyth, T. (1996) *Sustainable Tourism: Moving from Theory to Practice*, London: Tourism Concern, p. 31.
75   *Ibid.*, p. 6.

76 Brian Wheeller (1996) 'In Whose Interest?', *In Focus* 19, London Tourism Concern, p. 15.
77 Brian Wheeller (1994) 'A Carry-on up the Jungle', *Tourism Management*, 15, 3: 10.
78 *Ibid.*, p. 10.
79 Haslam, P. A. (2003) 'Voluntary Codes of Conduct in the Americas: Evaluating the State of the Art', Paper presented to the Wilton Park Conference on Business Ethics, UK, 16–18 October. Canadian Foundation for the Americas.
80 Peter Mason and Martin Mowforth (1995) 'Codes of Conduct in Tourism', Occasional Paper, University of Plymouth.
81 Naomi Klein (2001) *No Logo*, London: Flamingo, p. 434.
82 *Ibid.*, p. 437.
83 *Ibid.*, p. 430.
84 www.toinitiative.org.
85 More details of this best practice example can be found in Chapter 5 of the Case Studies section of the TOI website: www.toinitiative.org/good_practices/case_studies.htm.
86 As reported by Christine Kamp in the Clearinghouse service of the Tourism Investigation & Monitoring Team, 17 January 2004.
87 www.journeylatinamerica.co.uk/home/travellerscode.shtml.
88 February 2004.
89 World Tourism Organisation website, 'Global Code of Ethics for Tourism', www.world-tourism.org/quality.
90 Explore Worldwide website: www.exploreworldwide.com/worldwide/responsibletourism.jsp. Exodus website: www.exodus.co.uk/restourism.html.
91 Tearfund (2001) 'Tourism: Putting Ethics into Practice', Teddington: Tearfund.
92 André de Mendonca, personal communication, 5 January 2004.
93 Tearfund (2002) *Worlds Apart: A Call to Responsible Global Tourism*, Teddington: Tearfund, p. 10.
94 FTSE: Financial Times Stock Exchange. The FTSE is a trade mark of the UK Financial Times and the London Stock Exchange.
95 Tearfund (2002), p. 10.
96 Fair Trade in Tourism Network (2002) 'Corporate Futures: Social Responsibility in the Tourism Industry', London: Fair Trade in Tourism Network.
97 Haslam, P. A. (2003).
98 Fair Trade in Tourism Network (2002).
99 Haslam, P. A. (2003).
100 Tearfund (2002), p. 13.
101 *Ibid.*
102 Association of Independent Tour Operators website, 'Responsible Tourism Guidelines': www.aito.co.uk/home/responsibletourism.htm.
103 AITO (2002) 'The Independent Holiday Website', www.aito.co.uk.
104 AITO (1996) *The AITO Directory of Real Holidays*, London: AITO.
105 International Business Leader Forum (IBLF) website (2002) www.iblf.org.
106 Ashley, C., Roe, D. and Goodwin, H. (2001) *Pro-poor Tourism Strategies: Making Tourism Work for the Poor: A Review of Experience*, London: Overseas Development Institute.
107 Banana Link (2004) 'Editorial: WTO – Wiser Tariffs Opportunity?' *Banana Trade News Bulletin*, 30, April: 1.
108 Cleverdon, R. and Kalisch, A. (2000) 'Fair Trade in Tourism', *International Journal of Tourism Research* 2: 175.
109 *Ibid.*, p. 178.
110 Ashley, C. *et al.* (2001).
111 *Ibid.*, p. 41.

112 Hutnyk, J. (1996) *The Rumour of Calcutta: Tourism, Charity and the Poverty of Representation*, London: Zed Books, p. 215.

113 UK Department of Trade and Industry (2004) 'The Government's Expenditure Plans', White Paper on international trade and investment, 6 July 2004.

114 Bradshaw, S., Linneker, B. and Zúñiga, R. (2000) 'Social Roles and Spatial Relations of NGOs and Civil Society: Recent Participation and Effectiveness in the Development of Central America and the Middle America Region', in C. McIlwaine and K. Willis (eds) *Mexico, Central America and the Caribbean*, Harlow: Addison Wesley Longman.

3 Local politics, poverty and tourism: the micro level of responsibility

Tourism is often touted as a route out of poverty for those communities and nations which embrace it and promote it. It is particularly touted in this way by First World governments, companies and tourism organisations like the WTO/OMT and WTTC. Increasingly over recent years, First World development agencies and United Nations agencies have also come to accept that the industry can be seen as a vehicle for alleviating poverty. This may or may not be due to the failure (now obvious to some but not to all) of the neoliberal development model around the Third World. Three decades of following the prevailing dogmatic belief in comparative advantage, whose promoters, the IMF, WTO/OMC, World Bank and G8 governments, have forced Third World countries to produce tropical goods and to exploit their natural resources, have given rise to increases in poverty and inequalities. Hence the perceived need for change and the relatively recent use of the tourism industry as a 'new' mechanism either for continuing the exploitation or for addressing its failings. Hence too the recent rise to prominence of pro-poor tourism initiatives.

National governments in the Third World follow this line and view tourism as a means of increasing foreign earnings and as a generator of employment. There is also no doubt that many communities in the LAC countries view the industry with great hope.

This chapter considers the grounds for this hope and the expectations built upon it. It examines the effects of tourism on poverty and inequality and the likely effectiveness or otherwise of pro-poor tourism initiatives on the reduction or elimination of poverty, a goal to which many First World governments have rhetorically committed themselves. In order to make this examination, we first examine the nature of poverty and inequality in the LAC countries. If poverty is indeed to be reduced as a result of the operations of tourism, then it would seem important that the industry should incorporate the notions of participation, self-mobilisation and empowerment into its operations; and these ideas are also considered here. Again, as consistent with our approach in other chapters, we place these considerations in the context of the concepts of power and development. Do different understandings of development fit happily with the shift of power that might be required if self-mobilisation and empowerment become crucial to poverty reduction through tourism? The chapter also considers the much-used term 'community-based tourism' and seeks, briefly, to unpack the notion of community. As well as taking their own action, communities may receive the benefits and disbenefits of nearby tourism mega-developments. We consider whether poverty reduction can be brought about through such large-scale and imposed tourism projects which are often reported as offering only menial employment to local people rather than power and control to the poor.

There are two clearly different strategies at play here. The first is that of empowering local people and communities to take control and thereby to be the major recipients of the benefits of tourism activities. The second is that of dependence on outside agents to provide the necessary tourism infrastructure (such as hotels) and general infrastructure (such as roads and basic services), in the hope that the direct and indirect benefits of such investment will trickle down to the poor. We address the question of whether either or both of these strategies are likely to result in the stated goal of poverty reduction.

Poverty and inequality in the LAC countries and pro-poor tourism

Until the 1980s it was common to indicate the wealth of a country, and conversely its poverty, by the measure of its gross domestic product (GDP). The GDP is the total value of the output of goods and services produced by an economy, by both residents and non-residents.[1] It is used to show the relative wealth of different countries and through yearly comparisons to show levels of economic growth. The GDP and its associated measure, gross national product (GNP – see note 1), are still widely used. Economic growth, however, is not necessarily the same as economic health as Griffiths makes clear:

> El Salvador's macroeconomic figures for 1995 looked excellent . . . The country's 6% growth rate in 1995 led Central America and the Caribbean, and was only surpassed by Chile and Peru in Latin America . . . While financial experts lauded the figures as proof of El Salvador's success in implementing neoliberal economic stabilisation, statistics compiled by the ombudsman's office reveal how very little the macroeconomic surge has so far meant to many citizens left behind. 51% of Salvadorans live in absolute poverty, 40% of the population lacks access to health services, 53% have no potable water and 29% are illiterate, preventing them, tragically, from reading Milton Friedman.[2]

More recently (2002), the United Nations Economic Commission for Latin America and the Caribbean (ECLAC) has drawn attention to the fact that, despite a recovery in economic growth in the LAC countries between 1990 and 1997, since 1997 'the region has seen economic growth and the fight against poverty stagnate and, in some cases, clearly retreat', and 'today poverty is higher than in 1997 and . . . the number of poor rose by almost 18 million'.[3] Using data from the World Bank's World Development Reports,[4] in 2003 the *New Internationalist* magazine declared that on poverty 'no progress has been made since the 1980s in reducing

Table 3.1 Poverty indicators for selected LAC countries

Country	Human Development Index[1]	Per cent of population below income poverty line of US$1 per day 1990–2003[1]	Per cent of population below income poverty line of US$2 per day 1990–2003[1]	Tourism expenditure as a per cent of GDP 2004[2]
Argentina	0.863	3.3	14.3	2.0
Belize	0.753			12.2
Bolivia	0.687	14.4	34.3	3.0
Brazil	0.792	8.2	22.4	0.6
Chile	0.854	<2	9.6	1.7
Colombia	0.785	8.2	22.6	1.4
Costa Rica	0.838	2.0	9.5	8.6
Cuba	0.817			
Dominican Republic	0.749	<2	<2	17.0
Ecuador	0.759	17.7	40.8	1.2
El Salvador	0.722	31.1	58.0	4.0
Guatemala	0.663	16.0	37.4	2.9
Haiti	0.475			
Honduras	0.667	20.7	44.0	5.5
Jamaica	0.738	<2	13.3	21.6
Mexico	0.814	9.9	26.3	1.7
Nicaragua	0.690	45.1	79.9	4.4
Panama	0.804	7.2	17.6	6.5
Paraguay	0.755	16.4	33.2	1.2
Peru	0.762	18.1	37.7	1.7
Trinidad & Tobago	0.801	12.4	39.0	4.6
Uruguay	0.840	<2	3.9	4.4
Venezuela	0.772	15.0	32.0	0.5

Sources:
1 Human Development Report 2005 'International Cooperation at a Crossroads: Aid, Trade and Security in an Unequal World', United Nations Development Programme.
2 World Tourism Organisation (2006) *Compendium of Tourism Statistics: Data 2000–2004*, Madrid: WTO/OMT.

either the absolute or relative numbers of destitute people in Latin America'.[5] Table 3.1 gives United Nations poverty data for selected LAC countries along with data on expenditure by tourists as a proportion of each country's GDP.

Levels of poverty and of economic health can be shown in a number of ways and, as this brief discussion has made clear, no single indicator will suffice to demonstrate them accurately on its own. Poverty and economic health can be interpreted in both absolute and relative terms. Vandana Shiva makes clear that there is a distinction between poverty as subsistence and misery through scarcity and want:

> It is helpful to distinguish between a cultural concept of a simple and sustainable life, understood as poverty, from the material experience of poverty as a result of dispossession and scarcity. Poverty perceived as such from

a cultural perspective is not necessarily real material poverty: subsistence economies which satisfy basic needs by means of self-supply are not poor in the sense that they are wanting. The ideology of development, however, declares them to be poor for not participating significantly in the market economy and for not consuming goods produced in the global economy.[6]

Moreover, the United Nations Office of the High Commissioner for Human Rights (OHCHR) defines poverty as 'a human condition characterized by the sustained or chronic deprivation of the resources, capabilities, choices, security and power necessary for the enjoyment of an adequate standard of living and other civil, cultural, economic, political and social rights'.[7] This again emphasises that, although economic deprivation is an element of any definition of poverty, it is far from being the only element. So the measurement of poverty is far from an easy matter, and people's wellbeing cannot be measured simply by income or value of production. Wellbeing requires good health, access to satisfaction of the basic human needs, a safe environment and a strong communal life. Taken together, some of these measures are often used as indicators of a nation's level of development, and in 1990 the United Nations Development Programme (UNDP) published the first Human Development Report[8] which included data on the human development index (HDI) as a composite measure of human development.

By combining both social and economic indicators, the HDI attempts to reflect a general sense of wellbeing that people may or may not feel in their lives. The specific measures they combine are life expectancy, literacy and per capita income. The mathematical basis for, rational justification for, drawbacks of and results of the measure are given in much greater detail in various yearly editions of the Human Development Report of the UNDP.

Since 1990, other supplementary composite measures of development have been produced by the UNDP, one of these being the human poverty index (HPI). The HPI reflects levels of deprivation in a country through the same variables used in the calculation of the HDI. The HPI-1 measures poverty in developing countries through longevity (as the probability at birth of not surviving to age forty), knowledge (as the adult illiteracy rate), and overall economic provision (as the percentage of people not using improved water sources and the percentage of children under five who are underweight).

As the UNDP states '*all* definitions of poverty involve social judgements' (emphasis in original).[9] The UNDP defines *absolute* poverty by comparing personal or household income with the cost of buying a given quantity of goods or services, *relative* poverty by comparing that income with the incomes of others, and *subjective* poverty by comparing actual income against the income earner's expectations and perceptions. And as they add, 'There is no scientific, unequivocal definition of who is and is not poor'.[10]

Despite this definitional problem, measuring poverty and development is important and the use of composite indices, such as the HDI and HPI, is generally acknowledged to be more appropriate than single measures of income or value of productivity, such as GDP. But the HDI and HPI also fail to tell the whole story.

Also important, especially in the field of tourism, is income inequality within countries and regions, and, as Clifton Holland has pointed out, 'Latin America is one of the world's most unequal regions in terms of income distribution. This means that economic growth will not reduce poverty unless Latin American governments redirect it to the poor.'[11]

Inequality exacerbates the effects of market and policy failures on growth and thus on progress against poverty. The poor, for example, are generally unable to borrow as they have no collateral, and this impedes their ability to set up small businesses, as pertinent in the tourism industry as in others. Concentration of income at the top also undermines public policies, on education and health for instance, which will advance human development and the individual's potential to better themselves. Moreover, inequality erodes social capital and the ability of society to provide reliable institutions and services, and thereby renders the individual's trust in society less worthwhile, so that their participation loses significance.[12]

That inequality is important in the LAC countries is demonstrated by Oliver Marshall, co-author of the *Rough Guides* to Brazil: 'everywhere you go, extreme social and economic disparities are striking, nowhere more so than in the cities, where conspicuous wealth is displayed side by side with extreme poverty'.[13] The lesson from this is starkly put by José Antonio Ocampo, ECLAC Executive Secretary, when he states that:

> depending solely on economic growth to deal with the problem of poverty in Latin America will make it hard to meet the goal set for 2015. It is becoming ever more necessary to resort to economic policies that, aside from seeking to expand the productive base and increase national output, include the progressive redistribution of income as a viable alternative for meeting the Millennium targets.[14]

A country's levels of poverty, inequality and development matter primarily for the sake of that country's population. But they also matter to potential investors in the country, to potential visitors to the country and increasingly of late to the promoters of neoliberal economic development (the Washington Consensus). Poverty, inequality and a low level of development (Third World status) may in some cases serve as an attractor of tourists (see Figure 3.1); in other cases as a detractor of tourists. To members of the Washington Consensus (the World Bank, IMF, WTO/OMC and others identified in Chapter 2) high levels of poverty and low levels of development are a mark of failure of their model of economic development, one of whose principal tenets is that economic wellbeing will trickle down to the poor thereby spreading, eventually and theoretically, the benefits of this form of development to the whole population. If poverty is seen to increase and development is seen to stagnate or worse, then clearly the model is failing, even by its own criteria.

The need to examine the *modus operandi* of the prevailing economic model in order to reduce poverty and inequality significantly has recently been shown in the acknowledgement by the supranational organisations and international financial

Figure 3.1 Poverty and tourism

institutions of the need to promote poverty reduction. As Bridget Wooding and Richard Moseley-Williams point out: 'recognition of the links between wealth creation and poverty reduction in highly unequal societies has now led to national anti-poverty programmes in developing countries, backed by international financial institutions'.[15] The Millennium Development Goals are just one manifestation of this recent recognition. A number of developments in bilateral donor policies have also given cause for some optimism in this regard, among them the development of pro-poor tourism initiatives.

With the emergence of development approaches focused on the poor, by the end of the 1990s development practitioners had begun to think about the possibility of applying poverty elimination goals to tourism. Tourism had traditionally been viewed as a commercial sector, and not considered a serious development-related activity. Since the late 1990s, however, governmental development agencies such as the UK Department For International Development (DFID) have supported studies that seek to elaborate a pro-poor tourism approach and promote tourism as a 'legitimate' development activity. And in the first years of the new millennium, pro-poor tourism initiatives have become fashionable for all the supranational organisations involved in development.

The emergence of pro-poor tourism must be understood and analysed within the global development orthodoxy that economic growth is fundamental to pro-poor development. Advocates of pro-poor tourism point to four potential advantages of tourism as an economic sector capable of facilitating pro-poor growth: the high potential of linkage, labour intensity, tourism's potential in poor countries and the ability to build tourism on natural and cultural assets. Detractors are critical of tourism because of its level of leakage, negative impacts on the poor, displacement and socio-cultural disruption.[16]

Pro-poor tourism is defined by its proponents as tourism 'that generates net benefits for the poor . . . Rather than aiming to expand the size of the sector, pro-poor tourism strategies aim to unlock opportunities – for economic gain, other livelihood benefits, or engagement in decision-making – for the poor.'[17] Table 3.2 lists the main components of pro-poor tourism. Pro-poor tourism, it is argued, differs from other tourism types that claim to have some development value (principally ecotourism, sustainable tourism, community-based tourism, fair trade tourism and other ethically based tourisms), in that it focuses directly on the needs of the poor. However, advocates also recognise that general support for and integration of pro-poor strategies with the mainstream tourism sector (for example, developing tourist boards or the provision of key infrastructure such as roads or water supplies) is essential to complement pro-poor tourism, and that 'interventions do not always need to be poverty-focused to benefit the poor'.[18]

Despite the acknowledged desire of pro-poor tourism advocates to overcome existing problems and encourage new approaches to development through tourism, two key problems arise: the first relates to an understanding of power relationships in tourism; the second to the ability of pro-poor tourism to reduce poverty.

First, the DFID Report is critical of the misrepresentation of tourism as an 'industry where foreign interests dominate',[19] and it directly challenges critiques of tourism that have emphasised the significance of power relationships. However, the authors of the Report themselves point out that 'Developing countries have only a minority share of the international tourism market',[20] refer to the major players in tourism and their 'low level of commitment to any one destination'[21] and point

Table 3.2 Pro-poor tourism

Participation	Poor people must participate in tourism decisions if their livelihood priorities are to be reflected in the way tourism is developed
Holistic livelihoods approach	The range of livelihood concerns of the poor – economic, social, environmental, short-term and long-term – need to be recognised. Focusing simply on cash or jobs is inadequate
Distribution	Promoting pro-poor tourism requires some analysis of the distribution of benefits and costs – and how to influence them
Flexibility	Blueprint approaches are unlikely to maximise benefits to the poor. The pace or scale of development may need to be adapted
Commercial realism	Ways to enhance impacts on the poor within the constraints of commercial viability need to be sought
Learning	As much is untested, learning from experience is essential. Pro-poor tourism also needs to draw on lessons from poverty analysis, environmental management, good governance and small enterprise development

Source: Adapted from Ashley, C., Boyd, C. and Goodwin, H. (2000) 'Pro-Poor Tourism: Putting Poverty at the Heart of the Tourism Agenda', Natural Resource Perspectives No. 51, ODI, London.

to the significance of customer requirements – the fundamental needs and desires of tourists that must be met for tourism to be a going concern. And as later pro-poor tourism analysis suggests, 'Most of the critical decisions that affect the sector are made outside of the country or by a few powerful local interests'.[22]

All such elements seem to confirm, rather than question, the significance of the geographically uneven and unequal nature of tourism production, consumption and development. For these reasons we argue that tourism must be understood primarily within an analysis of relationships of power (foreign or otherwise) and that tourism must be considered little different from other economic sectors when assessed within a development context. Here we try to understand tourism within the context of power and to examine how this affects local livelihoods, and to measure the veracity of the claims that have been made about the ability of new forms of tourism to assist development, and in particular the ability of pro-poor tourism to reduce poverty.

Given the global consensus on the need to dramatically reduce absolute poverty, the second fundamental question centres on the ability of pro-poor tourism to deliver poverty reduction. As might be expected at the initial stage in the development of a prospective approach that is 'relatively untried and untested' and where a blueprint for implementation does not exist,[23] accounts of pro-poor tourism are replete with imponderables (the 'may', 'should', 'likely to') and with acknowledgement that impacts are not easy to measure objectively.[24] While this is certainly no good reason for not embarking on pro-poor tourism initiatives, it does perhaps call into question the confidence invested in pro-poor tourism approaches and the manner in which it has been preferenced and mainstreamed into bilateral development assistance.

The UN poverty reduction targets and the consensus surrounding the needs of the poor as *the* focus for development activities has meant that pro-poor tourism advocates have been careful to define their approach in contrast to other types of tourism, even where the needs of the poor are already being addressed. Inevitably, this has meant trying to rationalise away potential inconsistencies. So while the UN targets are unequivocally focused on the reduction of the poor living in absolute poverty, on less than a dollar a day, pro-poor tourism advocates point to the problems of defining poverty and the poor.

Ashley *et al.* (2001) have reviewed a number of pro-poor tourism case studies including two from Ecuador and St Lucia which are briefly described in Boxes 3.1 and 3.2.[25] They conclude that a 'focus on "the poor" often translates in practice to a focus on local residents or the "community"'.[26] From the same stable, namely the Pro-Poor Tourism Partnership,[27] Xavier Cattarinich has suggested that 'many actors pursue initiatives that have beneficial environmental impacts, but that those initiatives and impacts are of secondary importance for the poor'.[28]

As advocates of pro-poor tourism concur, it is certainly not a panacea: 'it is probably true that the small size of . . . initiatives has meant that tourism provides a *minor dent in the national poverty* even when multiplier effects are taken into account' (emphasis in original).[29] Pro-poor tourism is not, therefore, a tool for eliminating nor necessarily for alleviating absolute poverty, but rather a measure for making some sections of poorer communities better-off and of reducing the vulnerability of poorer groups to shocks (such as hunger).

Box 3.1   Tropic Ecological Adventures, Ecuador

Tropic Ecological Adventures is a small, for-profit company, driven by motivated individuals, with the specific objective of demonstrating the 'viability of environmentally, socially and culturally responsible tourism' as an alternative to oil extraction in the Ecuadorian Amazon. It leads tours to natural areas in Ecuador, including the Amazon, usually for small, high-paying groups. The company has gone beyond normal business practice of supporting community tourism to establish joint products with remote Amazonian communities. Tropic has worked with the Huaorani people bringing tourists into the community for overnight stays to experience Huaorani culture and lifestyle.

Tropic's customised tours give tourists the opportunity to experience the rainforest and have a respectful encounter with local, indigenous people. The company encourages tourists to engage in discussions with elders in a conscious effort to create balanced inter-cultural interactions. The market niche it has targeted is adventurous, relatively wealthy people (European and American) keen to explore the Amazon and learn from indigenous culture.

Income received by the community does not necessarily affect their level of subsistence, which remains still largely dependent on the forest, but can provide critical funds for education and both long-term and emergency health care. In Zabalo, the majority of households – especially those whose men work as either guides or cook-administrators – have been able to move up from the classification of 'poor' into a more stable economic condition. Among the Huaorani other community residents who have received less direct benefits from Tropic's initiative may still be considered 'poor' as they have not earned enough to have money consistently available. Residents in both communities have expressed opinions that selling handicrafts is one of the best parts of having tourists visit their communities. More than a few women commented that they want more groups to visit just so they can sell more handicrafts.

Although Tropic found that its community-based programmes were less profitable and less marketable than some of its other activities, it has overcome this problem by coupling them with more mainstream tours such as visits to the Galapagos Islands. A decline in tourism in the Ecuadorian Amazon in 1999 and 2000 due to kidnappings and political upheaval, however, drove down prices and undermined the company's impact-minimising approach.

Sources:
Caroline Ashley, Dilys Roe and Harold Goodwin (2001) *Pro-poor Tourism Strategies: Making Tourism Work for the Poor: A Review of Experience*, London: Overseas Development Institute.
Braman, S. and Fundación Acción Amazonia (2001) 'Practical Strategies for Pro-poor Tourism. Tropic Ecological Adventures – Ecuador'. PPT Working Paper No. 7. London: Overseas Development Institute.

Box 3.2   St Lucia Heritage Tourism Programme

The St Lucia Heritage Tourism Programme (HTP) is an attempt to shift a whole country's tourism sector on to a more sustainable footing. It refers not to a specific community, but is rather a country-wide initiative with a strong pro-poor component. The HTP's main objectives are to stimulate a wider distribution of the benefits of existing tourism (cruise ship passengers and staying visitors), to create a new tourism sub-sector, described as heritage tourism, and to encourage tourists to stay longer on the island and to spend more money in local communities as opposed to resorts.

The successes claimed by the programme include awareness raising, improving the profile of local operators, increased competition for clients on the cruise ship wharf, the development of new products and attracting tourists to inland initiatives. New products include turtle watching, a mangrove forest tour and a seafood night that attracts both St Lucians and tourists.

Problems encountered by the HTP include the dependence of poor people on communal assets, the lack of local capacity, the need for greater political support, the difficulty in attracting beach and package tourists away to cultural products and the slow pace of change brought about on the ground in such a comprehensive approach. Despite these problems it is believed that some shift in the entire industry towards more positive impacts on poverty and sustainability has been achieved.

Sources:
Caroline Ashley, Dilys Roe and Harold Goodwin (2001) *Pro-poor Tourism Strategies: Making Tourism Work for the Poor: A Review of Experience*, London: Overseas Development Institute.
Renard, Y. (2001) 'Practical Strategies for Pro-poor Tourism. A Case Study of the St. Lucia Heritage Tourism Programme'. PPT Working Paper No. 6. London: Overseas Development Institute.

But it is difficult to draw conclusions from national aggregate data and analysis that 'shows that in most countries with high levels of poverty, tourism is significant (contributing over 2 per cent of GDP or 5 per cent of exports) or growing (aggregate growth of over 50 per cent between 1990 and 1997)'.[30] While this may well present opportunities for the poor, we must also recognise the geographically uneven and unequal nature of tourism development and activity. As a prerequisite therefore, there needs to be a 'large number of poor people, in areas with tourism assets'[31] and the commitment of government to pro-poor approaches.

As tentative analysis suggests,[32] and as conjectured in pro-poor tourism analysis, 'Even if tourism does not directly involve poor people, it may have pro-poor impact, *if* it improves government revenue and *if* that revenue is used in pro-poor ways'.[33] This seems to suggest that members of local communities may derive benefit from their engagement with tourism without direct and proactive involvement in a specifically pro-poor tourism scheme. Perhaps this further suggests that a more productive means of reducing poverty through tourism would be to place more emphasis on the improvement of employment conditions (through minimum wage and working conditions legislation, health and safety issues, union representation and so on) in mainstream tourism activities than on the promotion of specifically pro-poor tourism initiatives.

Of course, there are advantages to the growth of new, alternative and responsible holidays (and pro-poor tourism) and there is a number of case studies that appear to demonstrate the real advantages and benefits to the poor that arise from tourism. As advocates of pro-poor tourism understandably maintain, while in absolute terms the scale of benefits may appear small, they can be relatively very significant when viewed from the perspective of the beneficiary groups. Our tentative conclusions, however, remain the same: given the growth in global inequality and poverty, the benefits to the poor of poverty elimination (however real they may be where they occur) will be marginal within the overall context of tourism.

We draw this conclusion partly because of the increasing significance of global and national influences on the life-decisions that people make at the local level, and the parallel decrease in access to power that local people have over the decisions which affect their lives. Box 3.3 illustrates the conflicts that can be caused by possible tourism developments in a region of El Salvador. The lack of access to power is the factor which prescribes the need to protest for the people concerned and reflects their lack of access to Salvadoran legislation (for land titling) and the struggle they have to wage for access to one of the basic human needs, namely land. It is this recognition of the importance of power structures together with the effects of economic globalisation at the local level which have forced an acceptance of the need not only for pro-poor tourism initiatives but also for community-based tourism initiatives. And it is to the notion of community-based tourism that we now turn our attention.

Community-based tourism

The holiday is a chance to suspend the routine of our daily lives in order to concentrate our thirst for enjoyment, titillation and entertainment into a week, a fortnight or a month of unfettered pleasure – a kind of binge-pleasuring. For many of us the last concern in such circumstances is the rules and regulations which in normal circumstances would restrain our behaviour. These too are suspended temporarily and, despite our willingness to part with our money at such times, our presence at the destination end is sometimes the cause of a number of headaches for members of the local community.

Quite apart from the problems associated with large numbers of visitors (extra litter, sewage, demand for resources, congestion, higher prices and so on), some tourist developments have led directly to displacements of local people from land and property that has been theirs for years and even generations. In First World nations, safe and sure tenure and land titling are generally taken for granted, whilst in the LAC countries land and property disputes, leading to evictions and

Box 3.3   Tourism-related land conflicts in the Lower Lempa valley, El Salvador

Throughout 2004, the Association of United Communities for the Economic and Social Development of the Lower Lempa in Jiquilisco (ACUDESBAL) and Voices on the Border, a Salvadoran NGO, closely followed debates in the Salvadoran National Assembly on the Law for Natural Protected Areas. Lower Lempa community leaders were fearful that the law could have been used to nationalise areas surrounding the salt mangrove forests that cover large areas of the Lower Lempa basin.

First drafts of the bill to pass the law contained language suggesting that land titles in areas 'previously covered by salt water forest vegetation' could be annulled. Aspects of the bill appeared to be aimed at evicting all or large numbers of inhabitants of the Lower Lempa and other coastal areas. Tourism plans for the salt forests of Jiquilisco Bay had been public knowledge for some time, and nullifying land titles in the area, as the bill proposed, would have allowed for land concessions to be granted by the Salvadoran government, especially to large investors in tourism for luxury hotels in the Bay area.

The 1992 Peace Accords (which brought an end to the twelve year war in El Salvador) included a Land Transfer Programme which legalised the transfer of most of the Lower Lempa lands to ex-combatants and refugees. Since that time, the Salvadoran government has repeatedly sought the removal of the inhabitants in order to return these lands to the wealthy families that had owned them prior to the war. Former President Armando Calderón Sol (1994–1999) tried to promote re-location in the wake of Hurricane Mitch. President Francisco Flores (1999–2004) launched a new effort in the aftermath of the 2001 earthquakes. In July 2003, during a discussion about the levées around the Lempa River, the Minister of Agriculture insisted to community leaders that their communities need to re-locate to other areas of the country. This led to demonstrations by communities on both sides of the Lempa in December 2003. Throughout 2004, communities in San Marcos Lempa cantón suffered government-sponsored land grabs, giving legal rights to property to campesinos from other parts of the country rather than to inhabitants of these areas who had been seeking legalisation of their land for years.

It is this context which fed and justified the fears that the wording of the proposed bill and its failure to recognise land rights of largely the poor in these areas were the latest attempt to re-possess these lands. Strangely – or perhaps not too strangely – the bill's proposals recognised individual land titles in other natural areas in which wealthy families have their country mansions such as Lake Coatepeque in Santa Ana.

On 30 September 2004, over 300 people (mostly from the Lempa River basin) protested outside the Legislative Assembly in San Salvador with a view to stimulating debate and awareness about this bill. Community leaders speculated that the bill would prepare the ground for transnational corporations not only to promote tourism enterprises but also to facilitate access to the local plant and animal life by pharmaceutical companies seeking monopoly patent ownership under the Intellectual Property component of CAFTA, the Central American Free Trade Agreement which became effective in 2006.

In 2005, amidst considerable public and media attention aroused by the protests, the Salvadoran legislature approved the Law for Natural Protected Areas with wording which allowed for land ownership of those with legal titles and also extended the definition by allowing for the proof of land ownership by 'testimony'. This represented a victory for the local communities and vindicated their prolonged and consistent protests. But they remain wary; they are conscious of the history of land possession and dispossession in their country and know that they will face further threats in the future, most likely from the tourism industry, the most recent fashionable industry to represent economic and cultural power.

Sources: Voices on the Border at: www.votb.org, and Geoff Herzog.

displacements, are common even when and where land titles have been granted. Many such evictions and displacements have been documented in the pages of *In Focus*, a journal produced by the UK-based Tourism Concern which campaigns for tourism that is just, sustainable and participatory.[34] In the literature most case studies of displacement and resettlement illustrate a deteriorating situation for those displaced. This has been especially well documented in cases where the development promotes mass forms of tourism, as in Guatemala for instance, where three hundred campesino families were evicted in June 1996 from land they claimed belonged to the state – police burned down their homes and arrested several of them – to make way for a Spanish businessman's plans to build a tourist complex.[35] But the *In Focus* journal[36] makes clear that displacement, often by violently enforced eviction, is also a feature of a surprising number of ecotourism projects (as opposed to mainstream tourism projects) in many parts of the world. The reader is again referred to Box 3.3 which illustrates one battle waged by a number of communities against the threat of displacement by tourism interests in El Salvador.

Increasing awareness of these problems has given rise to a form of tourism referred to as community tourism or community-based tourism, the essence of which is its intent that the benefits of tourism should go to the local community. Community-based tourism seeks to increase people's involvement in and ownership of tourism at the destination end. Ideally, control of the tourism operation or activity remains in the hands of some or all members of the local community. Probably the best place to start in an analysis of community-based tourism is the UK NGO Tourism Concern's *The Ethical Travel Guide* which is the altered title of the third edition of *The Community Tourism Guide*, the second edition of which was entitled *The Good Alternative Travel Guide*.[37]

The guide is intended for the increasing population of holidaymakers who have become aware of the problems caused by tourism in local communities at the destination end of tourism. It covers a range of possible holidays in over forty countries and in its second edition it outlines the idea of community tourism by explaining that 'you can help local people and still have a good holiday, simply by going on tours that involve local communities'.[38] Mark Mann (author of the second edition) continues by explaining that there is no rigid set of criteria for tours or destinations to qualify as community-based tourism, and that the guide's range of tours and destinations is very diverse. Without investigating it in depth, the second edition hints at one of the fundamental problems of community-based tourism, namely the notion of community itself:

> it's much harder to talk about 'communities' in Western countries, with their mobile and largely urban populations. Most of the tours and projects we list are rural. Although many people in the developing world now live in cities, the idea of community is hard to apply in an urban context. Few tourists want to spend their holiday in developing world cities anyway, so urban tours are pretty thin on the ground.[39]

We would disagree that the notion of community is hard to define in an urban context and take up this issue again in Chapter 6 where urban tourism in the LAC countries is discussed and illustrated in some depth. The idea of community in urban areas does exist, albeit in a form different from most rural communities. The community in central areas of Mexico City, for instance, is clearly very different from a Tzotzil-speaking village in the southern Mexican state of Chiapas.

In the urban setting, it might be generally assumed that the antithesis of external control in community tourism would be found in the operation of *favela* tours and their equivalent forms of tourism in other poor *barrios* or shanties of cities in the LAC countries. The *favela* tour is a relatively new form of tourism involving guided tours around generally poor areas of cities. By reputation *favelas*, or shanties, are the epitome of poor neighbourhoods, stereotypically associated with high levels of poverty, deprivation and crime, into which outsiders rarely venture without the security of an accompanying resident. It is in such places that urban gangs (the *maras* in Central America) operate, where drug dealers prey on the young and vulnerable, and where opportunities beyond prostitution and crime are limited. One would assume, therefore, that the new-style *favela* tour is more likely to be run by a local resident, wise to the ways of the *favela* and with suitable contacts who can offer protection against crime and violence, than by outsiders who represent commercial companies or non-governmental organisations. Deborah Dwek's work in Rio de Janeiro (which is featured in Chapter 6), however, suggests that the opposite may be true: 'while I enjoyed my tour round Rocinha with local resident and guide Luisa – the only working guide to have been trained by Reis and held up as the success story – the agenda on the tour was still very much that of the tour agency, the outsider'.[40] And in her conclusion, she opines that: 'while local people in Rocinha have their own ideas about how to exploit tourism, they do not appear to have a voice with which to express them or any power to effect change'.[41]

Mann recognises that there are difficulties in understanding what exactly a community is, but explains that for the purposes of Tourism Concern's guide, 'a "community" means a mutually supportive, geographically specific, social unit such as a village or tribe where people identify themselves as community members and where there is usually some form of communal decision-making'.[42]

Although a community can be defined by scale, sector, interest, level of power, location and by numerous other features, its common usage, as illustrated by Mann, implies a homogeneity of membership that may not exist in reality. In other words, it is a common and rarely questioned assumption that when we talk of a community we are talking of all its members, as if they are all of one mind, homogeneous and static. It is a reasonable contention, however, that a majority of communities include a wide range of people with a wide range of opinions and perspectives. Moreover, this range of perspectives is likely to reflect differences in needs and interests which may in turn reveal different levels of power. These differences may also correlate or be associated with differences in race, class, sex, age or other attribute, and may be random or systematic. Even in tribal communities, whose members are strongly bound by communal norms, these differences exist.

It is through these differences that local elites are formed, and it is through these that power is exercised differentially and preferentially, leading to the bestowing of favours, resources and benefits on to selected members or sub-groups of the community. The Kuna Indians of Panama, who are the subject of one of the case studies given in Chapter 5 (pages 152–5), are a case in point. In the main group of the San Blas Archipelago of islands on which the Kuna live (off the Caribbean coast of Panama) there are only three hotels. It is generally perceived by the Kuna, for a number of reasons, as being in their own interests to keep it this way in order to prevent too great a number of tourists visiting their islands and in their attempts to prevent foreign involvement in their islands' tourism industry.[43] But many of the Kuna manage to derive financial benefit from the tourists by selling their appliqué cloths (*molas*) and other craft goods. The greatest profits from tourism, however, undoubtedly accrue to the owners of the hotels and their families. Doubtless also these families gain respect from this position and in turn derive more than average influence in the development of their islands. In the case of the Kuna, this is not an influence which is generally begrudged on account of the general benefit derived from tourism and the decision-making systems in the community. Regardless of how this power and influence is used, however, the case serves to illustrate how local elites may be formed through the tourism industry.

Powerful elites, however, are not immune to the changes brought by time. Some community members may be catalysed by contact with and involvement in earlier phases of tourism. They thus acquire a competence (in terms perhaps of entrepreneurial skills) and possibly resources unavailable to others in the community which enable them to assume a less dependent and relatively more powerful role in the tourism system at least within, if not beyond, the community.

We are in agreement with Mark Mann, Tourism Concern and other proponents of community-based tourism that it is a sound idea in principle. We raise our points of question only in an attempt to improve its practice and to raise general awareness of the issues behind community-based tourism. The Tourism Concern guide is an important possession for holidaymakers who want to exercise their tourism in a responsible manner, and as more communities try to control their own tourism developments and as awareness of these spreads, future editions can only enhance its usefulness.

To some extent, the development of community-based tourism followed on from the earlier form of tourism known as solidarity tourism. This was of some significance wherever liberation struggles were taking place, and in the 1970s and 1980s Central America was the crucible for a number of such wars against repressive regimes supported by a US foreign policy which favoured right-wing authoritarianism over struggles for social justice. In the 1980s, whilst US-financed, trained, armed and organised counter-revolutionaries (known as the contras) waged war against the left-wing Sandinista government of Nicaragua, the UK-based Nicaragua Solidarity Campaign organised solidarity tours of Nicaragua. Similar organisations in the USA led similar tours to Nicaragua; and Cuba, Chile, El Salvador and Guatemala all had their solidarity support groups in western countries, many of these running occasional 'revolutionary' or 'solidarity' tours.

Members of solidarity tour groups were generally young and politically focused on opposition to an aggressive, arrogant and interventionist US foreign policy. They helped with the coffee harvest, built and painted schools and health centres, and lived in rural communities amongst those they were helping. Throughout the 1990s, as the liberation struggles drew to an end, as membership of such tours lost its chic and as the hegemony of western capitalism lost its only serious challenger, solidarity tours slowly metamorphosed into political tours focused on the effects of globalisation, gender issues, trades unionism, fair trade or environmentalism. They remain a tiny sector of the tourism market, but it is interesting that they have not disappeared; and indeed, some of the old-style solidarity tours still exist, as can be seen in Box 3.4, which gives extracts from an advertising leaflet for a two-week ethical/eco visit to the community of Nueva Esperanza in El Salvador in 2005. It is clear from Box 3.4 that the community element of the tour is especially strong.

In one particular case, the passage of time has seen the return of revolutionary chic. In Bolivia in October 2004, the Che Guevara Trail was opened with a view to giving a boost to the country's flagging tourism industry and to stimulating small-scale, local enterprise along the trail, and especially in the community of La Higuera. Ernesto Che Guevara, hero of the Cuban revolution and revolutionary icon elsewhere in the world, was executed by Bolivian troops in 1967 in La Higuera, a remote village in Bolivia's eastern lowlands. He is still remembered by people in the community of La Higuera, which hosts a large bust of Che erected in 1997 to mark the thirtieth anniversary of his death. Julia Cortes, a nineteen-year-old trainee

Box 3.4   Solidarity tour to El Salvador, 2005

'El Salvador for Beginners' 2 weeks Ethical/Eco Visit Easter 2005

Bienvenidos! Bienvenidas!

Welcome to our community!

The ex-refugee community of Nueva Esperanza ('New Hope') INVITE YOU for an amazing 2 week experience:

- Ethical – you help fund local sustainable development
- Eco – you get to explore the stunning rural environment

Organised by Nueva Esperanza Support Group, Birmingham (0121 523 4118)

Total cost £1000

– all inclusive and all inspiring

What will the trip involve?

- Stay in basic but clean accommodation in the community's guest house
- Three delicious organic meals a day – vegetarian or vegan by arrangement
- Trips organised to local beauty spots
- Visit local projects according to your interests
- Meet and talk with ex-refugees and ex-combatants from the civil war (1980–1992)
- Experience the vibrant cultural life of this community

What is included in 'all inclusive'?

- Return flight via the USA
- Transport to and from the airport
- 13 nights accommodation and full board
- 5 or so trips out of the community
- Evening socialising within the community (but not the cold beers!)
- Services of a guide/translator/facilitator

Ethical – the community benefit directly from your visit, on their terms

Eco – an amazing natural tropical environment, bustling with small-scale progressive rural development projects

Source: Reproduced by kind permission of Nueva Esperanza Support Group, Birmingham.

teacher at the time of his execution, was one of the last people to see him alive. 'When I met him, he struck me as a person blessed with great charisma and intelligence,' she says. 'I brought him soup and we talked; he was very polite and respectful to me. Not at all dangerous.'[44] As well as the people of La Higuera, beneficiaries of the initiative will include some of the Guarani indigenous families who live along the trail in one of the poorest areas of Bolivia.

Preparatory work for the inauguration of the trail took three years and was carried out by CARE Bolivia, the local branch of the INGO CARE International, with funding from the UK Department for International Development (DFID). Jacqueline Peña y Lillo, project manager for CARE Bolivia, explains: 'The objective is not to exploit Che's name but to help local families through the creation of small-scale, tourist-based enterprises.'[45] David Atkinson further explains: 'Local families are being employed in new cultural projects, improving the services available to tourists and as official Che trail guides. As the project grows, CARE Bolivia aims to hand over its management entirely to the local community.'[46] Anita Pleumarom, a critic of such ventures, offers a comprehensive swipe at the project, its context, the motives of its initiators and at the industry in general:

> It appears to be an odd fate for the legendary revolutionary who had come to Bolivia to empower the rural poor to improve their lives and paid with his own life. After 37 years, the living conditions for the rural population have not much improved. But tourism consultants and managers are moving in to 'save' the communities in need. And because it is nowadays so easy to sell tourism-for-profit activities as altruistic 'poverty reduction' projects, it is not surprising that the 'Che tourism' venture is run by an NGO – the local branch of the UK-based international relief agency CARE – and partly funded by the British DFID.[47]

At the point of publication it is still too early in this case to discern whether the intention of transferring control from an outside agency to the local community will fully materialise.

To some extent the old-style solidarity tour, targeted especially at the young, has also been overtaken by the amazing growth and rise to fashion of gap year travel. This involves particularly student-aged youths keen to learn about the world beyond their home district, keen to avoid or delay their entry into the world of work and careers, and keen to tap the financial resources of their parents to the greatest possible extent. Their travel is often justified by the inclusion of volunteer work into their programme and this normally involves assistance with community projects such as building, painting, repairing, tree planting, trail clear-ups, environmental education and many other activities. It stretches the definition and the imagination to refer to it as community tourism, however, and so this subject is briefly covered more appropriately in Chapter 8.

Whilst the Tourism Concern guide offers no shortage of good examples of community-based tourism, beyond the guide there is also no shortage of tourism development case studies where critics have justifiably pointed out that control rests outside the destination community involved. In fact where local communities are supposed to be involved in a tourism development, their involvement is often more a shade of participation rather than a matter of control. This debate about degree of control is thrown into sharp contrast by the two standpoints of 'host' communities as objects of tourism or as controllers of tourism, a debate often considered to be at the heart of the difference between conventional mass tourism and supposedly sustainable and responsible new forms of tourism. We would argue, however, that the issue of control is the same whether it refers to mass tourism or any of the new forms of tourism, including community-based tourism. The issue of participation, control and their different shades are considered further in the following sub-section.

An observation regarding the role of outsiders in community-based tourism is made by James Fair, who asks if ecotourism can save Ecuador's threatened cloud forests.[48] He describes attempts to stimulate tourism that will benefit local residents in the Bellavista private cloud forest reserve owned by a British teacher and ecologist. His second example is another area of land bought by the Fundación Maquipucuna, admittedly an Ecuadorean foundation but one that is still seen as an external agency by local residents in the area. Other examples he cites involve the purchase of land in Ecuador with the help of UK-based organisations and the management of nature-based tourism schemes. Whilst Fair discusses the drawbacks of these schemes – the prohibition of hunting for food, the cause of divisions and envy between those who benefit and those who don't – as well as their successes in conservation and relatively insignificant creation of employment and economic opportunities, he also notes that 'local community initiatives involving eco-tourism in the Ecuadorean highlands are notable by their absence', and concludes that 'ecotourism, by its very nature, is more likely to be an idea conceived by outsiders'.[49]

As illustrated by the examples already given in this chapter, community-based tourism has been adopted by NGOs and INGOs as one form of development that is generally benevolent and which might have the potential for economic advancement of at least a few members of a given community. There is, however, a danger that local communities become commodified by the many conservation and social NGOs that seek to utilise local communities in order to secure grants and aid. Anita Pleumarom wonders 'whether the commercialisation of virtually everything that characterises community life and local culture can be considered more acceptable or ethical when the marketing is handled by NGOs in the name of "community-based tourism" for poverty alleviation'.[50]

Some of the large international organisations, such as Conservation International and The International Ecotourism Society, have produced manuals for local communities and others on how to practise ecotourism responsibly and how to make the prevailing system work for them. Such publications effectively explain how to work within the system of current power relationships to promote their own project or development. Some critics argue that this only strengthens and perpetuates existing power relationships rather than bringing about change; others argue that it is only in this way that the system will change, even if only incrementally.

Why it should be the case that such an apparently high proportion of small-scale, local, supposedly responsible tourism schemes in the LAC countries are begun, owned and run by outsiders, notably North American or European individuals or NGOs, may not be too clear. It is clear, however, to anyone who travels around such schemes in the LAC countries, physically or electronically, that it is so. Indeed, in the case of the Toledo Ecotourism Association in the south of Belize (featured as a case study in Chapter 5), one volunteer worker notes that 'the tourist board does not advertise the village guesthouses properly precisely because they are community-based rather than developed by the government or foreign investors!'[51]

As we jet around on our tourist circuits, it is difficult to escape the conclusion that in developing countries even community-based tourism is an industry for foreigners run by foreigners. In part, of course, this is explained by the need for tourism infrastructure investment which poor communities lacking access to power, capital, land and resources cannot provide. In part also it may be explained by the expectations of the clientele, knowledge of which is much greater amongst outside agencies than it is amongst local people who live in poverty with few if any of the services which the visitors take for granted.

It is possible, then, that there is more scope for substantial involvement of local communities in catering for the domestic tourism market rather than the international tourism market. Domestic tourists may expect different standards of provision from those that can readily be supplied by the poorer, small-scale suppliers. David Barkin suggests this in his observations of tourism based on the Monarch butterfly zone in Michoacan, Mexico,[52] which are highlighted in Chapter 4. He is critical of the dependence on foreign tourists brought in by tour operators based elsewhere in Mexico and the disruption they cause as well as the dependence on a short season. He claims that rural communities might benefit more from offering broader-based rural recreation and accommodation to middle- and lower-middle-income Mexicans. They would know the expectations of their clients better, their catering preferences and their language and may therefore be somewhat shielded from the volatility of the international market. In the short term it may be less sexy and less rewarding, but in the long term it may be more reliable.

But, as James Fair puts it, the conclusion that community tourism is run by outside agents 'does not mean that local people are not involved, nor that they could not one day take greater control over such projects'.[53] This may seem rather vainly hopeful, but around the subcontinent there do exist many projects and destinations where local people are in control of their own tourism services. One such is the case of Finca Sonador in the south of Costa Rica, details of whose community-based ecotourism scheme are given in Box 3.5.

The establishment of the community of Finca Sonador was assisted in the first place by the Longo Maï movement and the then president of Costa Rica, Rodrigo Carazo. Despite the continued residence in the community of a Swiss person originally associated with the Longo Maï movement, the community's ecotourism developments are an example of a local initiative with all the tourists' activities under the direct control of the community. There is very deliberately no outside

Box 3.5   Finca Sonador, Costa Rica

The Longo Maï movement, based in France, aims to give war refugees a positive and productive home and work environment rather than a temporary and transient camp. In 1978 the movement helped a group of Nicaraguan refugees fleeing the dictator Somoza's terror to form a small community in southern Costa Rica. After Somoza's overthrow in 1979 they returned to Nicaragua but were soon replaced by Salvadoran refugees fleeing the same type of state terror.

The community is now called Longo Maï – although some call it Finca Sonador – and has a population of around 400, mostly Salvadorans with a few Costa Rican campesino families, refugees from poverty and landlessness in their own country. The Finca is a relatively self-sufficient agricultural village which produces coffee and a few other products for sale beyond the village. In low-income months their traditional survival agriculture is based on corn, beans, pumpkins, rice and yucca (casava). Since 1992, the Finca has diversified its economic activities by attracting visitors. About 40 families are willing to accommodate visitors and have the space to do so. The *Comité de Turismo* ensures that all associated families receive visitors at times and that prices are uniform.

Prices for meals and accommodation (2007) vary between 8 and 12 US dollars depending on the length of stay. Most visitors are students doing projects and staying for weeks and often months. By the standard of the northern professionals, prices are ridiculously low. Other activities offered in the village include horseriding, guided tours, involvement in farming and general inclusion in communal and family life. Project work, donations (particularly for scholarships) and a small 'tourism tax' to co-finance communal projects through the *Comité de Turismo* ensure that the tourist money is distributed further afield within the community. Advertising for the scheme is largely by word of mouth, although a publicity sheet is posted in the Quaker lodging house in San José, and the community also has a website: www.sonador.info/

The small scale of the scheme would seem to be an essential feature both for the tourist and the host. For the visitor it is important that the experience has its air of exclusivity in the sense that this is not the usual tourist experience. For the host, it earns a little extra income with little extra cost and does not disrupt the community's or the family's way of life.

Sources:

Mowforth, M. (1996), 'Co-operativo Longo Maï, Costa Rica', *Newsletter of the* • *Environmental Network for Central America* 19: 6–7.

Finca Sonador (2000) 'Volunteer Work in Longo Maï and UNAPROA, Costa Rica', Finca Sonador information sheet.

Jovino Pérez and Liz Richmond (2003) 'Ecotourism Keeps Coffee Farmers Afloat in Costa Rica', *Newsletter of the Environmental Network for Central America*, 34: 9–10.

Cristoph Burkardt (2007) Personal communication

control – no tour operators, travel agents, local government officials, Institute of Tourism officials, and not even any *transportistas*. Visitors arrive there under their own steam or by public transport and walking, and advertising of the scheme is by word of mouth (with the one exception of an information sheet posted on a notice board in the Quaker lodging house in the capital city, San José). The scheme is also now listed in the third edition of Tourism Concern's *The Ethical Travel Guide*, but community representatives have asked for no entry or publicity in the mainstream travel guides. Finca Sonador illustrates one of the advantages of community-based tourism in that money paid for services rendered goes directly to those who render them without being 'creamed off' or cut down to a minimum by middlemen and agents.

The Finca Sonador example is something of an exception, but other exceptions exist in the subcontinent and many of them are listed in the Tourism Concern guide. The exceptions as well as those examples which seem to fit the general rule of control by outsiders rather suggest that it is not easy for community-based tourism schemes to serve the mass market and their large-scale tourism supply chains. It is possible that the hurdles to overcome in order to participate in the many stages of tourism supply chains are simply too high. The same may also be true of pro-poor tourism initiatives in which a number of tourism operations (such as car hire, coach operation, the provision of air-conditioned accommodation and so on) will normally be out of the reach of the local and poorly capitalised entrepreneur.

In the example of the Kuna Indians given above and in many other case studies given in this book, it is difficult not to see the crucial differences in levels of participation between different groups of people affected by tourism developments. It is because of the importance accorded to participation in almost all definitions of sustainability and responsibility in tourism developments that we now examine the notion.

Local participation in tourism developments

In recent years, there have been an increasing number of comparative studies of development projects showing that 'participation' is one of the critical components of success . . . As a result, the terms 'people participation' and 'popular participation' are now part of the normal language of many development agencies, including non-governmental organisations, government departments and banks. It is such a fashion that almost everyone says that participation is part of their work.[54]

One of the criteria often agreed as essential to the conditions of sustainability, responsibility and development in any new tourist scheme is the participation of local people. For the most part there has been an overwhelming benevolence towards the process of participation, and the activity has become mainstreamed in the work of many INGOs, multilateral agencies[55] and bilateral agencies. Indeed, the 1990s was the decade of participatory development. As Henkel and Stirrat argue, 'It is now difficult to find a development project that does not . . . claim to adopt a "participatory" approach involving "bottom-up" planning, acknowledging the importance of "indigenous" knowledge and claiming to "empower" local people'.[56]

Numerous community-based tourism schemes depend on the willingness of community members to participate in the scheme. Those members and families who participate will inevitably be those who derive economic benefit from the scheme. On account of this experience, they are also likely to be those who will be invited to participate in future such schemes in the community, and in this way their favoured position in the community will be consolidated over time. An understandable first reaction to this situation is to declare its soundness and to commend and support the willing participants. But in the same way that neoliberal development falsely assumes a level playing field on the global stage, this erroneously assumes a level playing field on the local stage. It assumes that members of a community are willing and able to participate equally, which in turn is based on the assumption that communities are the 'natural social entity',[57] an identifiable reality. Moreover, it assumes away the unequal access to power, the existence of conflict and exclusion within communities, and the social structures that influence the conduct and outcome of participatory processes.

Participation is not without its critics.[58] Cooke and Kothari, for example, refer to participation as the 'new tyranny', and make a critical attack on much development practice. Theirs is also a critique which seeks to expose and understand the sanctity in which participation is held, and the manner in which there are at times 'evangelical promises of salvation';[59] and there is a whiff of spiritualism in participatory practices. Thus, phrases such as 'targeting local people' and 'eliciting community-based participation'[60] and sentiments such as 'environmentally sustainable development . . . rests on gaining local support for the project'[61] and 'projects must provide direct benefits to local peoples'[62] come from the perspective of the project planner, usually from the First World, as are all these examples. The planners are often associated with a major INGO (such as WWF, Conservation International and The International Ecotourism Society as in these cases) or a supranational institution such as the World Bank (as in two of these cases) and all seek their own form of sustainability and responsibility through their appropriate projects.

It is not so much the good intentions or ethical and theoretical value that lie behind participation that are open to question but, rather, the often uncritical manner in which participation is conceptualised and practised that has drawn increasing attention.

Cleaver[63] argues that a new faith in participation arises from three key tenets: that participation is inherently good, that good techniques can ensure success, and that considerations of structures of power (and politics) should be avoided.

Regarding its inherent goodness, requiring an 'act of faith in development',[64] far from the exercise of a value-free approach, Henkel and Stirrat suggest that what the 'new orthodoxy boldly calls "empowerment"' has special resonance in what Michel Foucault calls 'subjection',[65] where the technical framework, approach and means of participation in participatory rural appraisal (PRA) is preordained and fixed. Ultimately, critics argue, this form of participation drives participants to seeing and representing their world within the context of the PRA 'expert's' vision. Or perhaps, local people are simply pragmatic and are able to offload local knowledge into predetermined structures, but with the view to realising opportunities and resources from external programmes.

Regarding the success ensured by good techniques, there is an underlying assumption that participation is a trip-switch to development. Kothari argues that PRA and similar participatory techniques represent an act with participants performing distinct, 'contrived' roles and practitioners or facilitators acting as 'stage managers or directors who guide, and attempt to delimit' the performance of participants.[66] In this way participants provide the information they believe is required to secure support;[67] as Cooke suggests, 'participatory processes may lead a group to say what it is they think you and everyone else want to hear, rather than what they truly believe'.[68] And as Mosse argues, rather than 'local knowledge' structuring and modifying development projects, formulaic project frameworks (such as widely used logical frameworks) relying on participatory planning techniques, in effect structure and articulate the local knowledge.[69] As such there is evidence that development practice is increasingly influenced by western managerialist thinking.[70]

Regarding the avoidance of consideration of structures of power and politics, Cleaver forcefully argues that an emphasis on perfecting method has inevitably resulted in a belief in problem solving through participation, but that this belief fails to acknowledge the structures of power, both within communities and between these communities and outsiders conducting participatory exercises. This aspect is especially significant within the overall context of this book, with its emphasis on relations of power.

The three major problems are the notion of a community as the '"natural" social entity'[71] and identifiable reality, the manner in which the heterogeneity and unequal access to power are assumed away, and the assumption that members of a community are willing and able to participate equally. The latter has been an enduring debate and problem within community development studies. The emphasis on solidarity in communities together with a closed and bounded conceptualisation of place, culture and community[72] leads to the relegation of conflict and exclusion in communities, and a failure to understand social and power structures that greatly influence the conduct and outcome of participatory processes. In Taylor's view, participation is simply not working, because it has been promoted by the powerful, and is largely cosmetic, but most ominously because 'it is used as a "hegemonic" device to secure compliance to, and control by, existing power structures'.[73] As such, then, participation simultaneously veils and legitimises existing structures of power.

While participation is a fundamental means of interaction and 'development', it is certainly not a panacea and does not automatically or necessarily lead to a change in the underlying structures of power. There are many well-documented examples of the relative lack of power held by local people in tourism developments in their locality – Brandon cites over fifty schemes, 'many of [which] had initiated nature tourism activities, but few of the benefits went to local people'.[74] (See also Johnston,[75] Wells and Brandon[76] and West and Brechin.[77]) This exclusion of local people from involvement and decision-making in the operation and benefits of tourism can be seen in some of the examples cited in this chapter and elsewhere in this book. It also provides an important justification for a differentiation between levels of participation, acknowledgement of the existence of which may help to move us away from the simplistic assumption that participation is of only one nature.

Pretty's typology of participation

The principle of local participation may be easy to promote; the practice is more complex, and clearly participation may be implemented in a number of different ways. Jules Pretty[78] has identified and described different types of participation as shown in Table 3.3, which offers a critique of each type.

Local circumstances, the unequal distribution of power between local and other interest groups, and differing interpretations of the term 'participation' are reflected in Pretty's typology of participation, which is just as applicable to the idea of 'partnerships', another mantra of the current phase of development, as it is to the idea of participation. Pretty's typology is especially helpful in developing an understanding of the factors which affect the development of tourism schemes in

Table 3.3 Pretty's typology of participation

Typology	Characteristics of each type
1 Passive participation	People participate by being told what has been decided or has already happened. Information being shared belongs only to external professionals.
2 Participation by consultation	People participate by being consulted or by answering questions. Process does not concede any share in decision-making, and professionals are under no obligation to take on board people's views.
3 Bought participation	People participate in return for food, cash or other material incentives. Local people have no stake in prolonging technologies or practices when the incentives end.
4 Functional participation	Participation seen by external agencies as a means to achieve their goals, especially reduced costs. People participate by forming groups to meet predetermined objectives.
5 Interactive participation	People participate in joint analysis, development of action plans and formation or strengthening of local groups or institutions. Learning methodologies used to seek multiple perspectives and groups determine how available resources are used.
6 Self-mobilisation and connectedness	People participate by taking initiatives independently of external institutions to change systems. They develop contacts with external institutions for resources and technical advice they need, but retain control over resource use.

Sources:
Pretty and Hine 1999; adapted from Pretty 1995. Jules Pretty (1995) 'The Many Interpretations of Participation', *In Focus* 16: 4–5, London: Tourism Concern. Pretty, J. and Hine, R. (1999) *Participatory Appraisal for Community Assessment: Principles and Methods*, Centre for Environment and Society, University of Essex.

local communities, and the remaining case studies illustrated in this chapter are referred to the typology.

The six types of participation range from *passive participation*, in which virtually all the power and control over the development or proposal lie with people or groups outside the local community, to *self-mobilisation*, in which the power and control over all aspects of the development rest squarely with the local community. The latter type does not rule out the involvement of external bodies or assistants or consultants, but they are present only as enablers rather than as directors and controllers of the development. The range of types allows for differing degrees of external involvement and local control, and reflects the power relationships between all players in the activity. For local people, involvement in the decision-making process is a feature of only the *interactive participation* and *self-mobilisation* types, while in the *functional participation* type most of the major decisions have been made before they are taken to the local community. The only forms of local participation that are likely to break the existing patterns of power and unequal development are those which originate from within the local communities themselves. This chapter provides a few such examples, but even these illustrate the fact that local circumstances always manage to complicate the best of intentions and the best of practices.

It would be easy here to make the prescriptive assumption that the greater the degree of local participation, the better the project. There are those, however, who might disagree with this assumption, especially, but not exclusively, those who represent a vested interest in a particular development project – the development agencies, governments, supranational institutions, or operators for instance. In these cases, some of the lesser types of local participation might be considered preferable to such external agencies. It is precisely this point which emphasises the importance of the power relationships involved in any tourist development project, and the fact that Pretty's typology reflects this underlines its value.

At this point, it is worth contrasting a number of examples of local participation in tourism developments in order to illustrate the manifestations and effects of different levels of involvement. We have attempted simply to describe the situations of each case study and to relate it to Pretty's typology, which allows us to make a consideration of the power vested in each interest group and their relation to the local community.

Box 3.6 includes excerpts from an article by Phil Gunson on the large-scale Mundo Maya (Maya World) project which covers five southern Mexican states plus Guatemala, Belize, El Salvador and Honduras. It is described in Mundo Maya publicity material as 'a ground-breaking tourism and regional development initiative . . . [which] seeks to improve the lot of area inhabitants with low impact projects which give visitors the opportunity to explore the area'.[79] In 1991 the project initially received US$1 million from the European Commission to promote three kinds of tourism in each country: cultural tourism, coastal tourism and eco/adventure tourism. The project promotes infrastructural improvements, new hotel construction, archaeological projects and extensive international marketing through glossy brochures, in-flight magazines and travel trade shows.

As Box 3.6 makes clear, there appears to be little or no attempt to involve local communities in decision-making. As the editor of *Tourism Link* (a journal of the Belize Tourism Industry Association) explained, 'full decision making powers for all Mundo Maya affairs lie in the hands of only five persons – basically the top public sector tourism officials of each country'.[80] The fact that this statement came as part of an article of complaint by private sector representatives about public sector control of the project underlines the irrelevance of local communities in this contest for power. According to Pretty's typology, this example might be classified as *passive participation*.

The Las Terrazas community in the Pinar Del Río province of Cuba, about 75 kilometres west of Havana, is located in a narrow valley reforested by the revolutionary government in the 1960s. The adjacent Sierra del Rosario became a UNESCO designated biosphere reserve in 1985, and in 1994 the Moka Ecolodge (shown in Figure 3.2) was built between the community and the sierra with Cuban government funds. The then Minister of Tourism, Osmany Cienfuegos, explained that 'What we've tried [to do] is to incorporate the natural environment and the

Box 3.6   Local participation in the Mundo Maya scheme

Marketing men put curse of tourism industry on Mayas

Order a prawn cocktail in a hotel in Chetumal, south-east Mexico, and it will probably come smothered in 'Mayan sauce'. A trivial example, but one that shows how the tourist industry, helped by Latin American governments, is turning a great pre-Columbian civilisation and its present-day descendants into a marketing concept.

But critics, including Mayan organisations, claim that archaeological sites and indian villages face being turned into a giant theme park, and that the millions of indigenous inhabitants have no part in decision-making. 'The bottom line is that they are just exploiting the resources of our people', says Greg Cho'c of the Kekchi council of Belize. 'Mayan people are not involved and cannot influence the project.'

The aims of the Mayan World scheme . . . include improving the quality of life of local residents, protecting the environment, and safeguarding historical and cultural heritage. But the Mexican government's own archaeological and cultural institute, the INAH, is sceptical. 'They have no awareness of what ecology is', says the director of the local INAH office, Adriana Velásquez. 'If they put up a palm-thatched hut they think it's "ecological".'

She cites the once-unspoilt Xcaret ruins, which have been turned into a park for day-trippers from the up-market resort of Cancún. The entrance fee is about £13, out of the reach of local people.

Rolando Pérez, a Quiché Maya, is one of about 30,000 Guatemalan refugees living in south-east Mexico . . . Mr. Pérez . . . believes white and mixed-race people want to eliminate the indians. 'They see us as an obstacle to development', he says. 'They just want to build big hotels for the tourists. They're the ones that benefit, not us.'

Local initiatives, such as a village guesthouse scheme started by Mayan villagers in Belize, have been ignored, says Stewart Krohn, managing director of Channel 5 television in Belize . . . 'If you go to a meeting of the Mundo Maya you won't find a Maya there, except maybe serving dinner', Mr. Krohn says. 'The Mayan people are just being used as low-cost labour. If I was a Maya, I'd put sugar in their gas tank.'

Source: Phil Gunson (1996) 'Marketing Men Put Curse of Tourism Industry on Mayas', London: *Guardian*, 28 September, p. 14.

Figure 3.2 Hotel Moka, Las Terrazas, Cuba
Photo: Martin Mowforth

local community. The idea is that the tourists and the community together participate in all this.'[81] Words like these have been echoed by many Tourism Ministers throughout the LAC countries about myriad schemes in the subcontinent, and we would be wise to deploy our natural scepticism and ask ourselves whether the sentiments expressed by Cienfuegos ever really materialised. But the exercise in participation conducted in this case might serve as something of an exemplar to all the report-writers and development workers whose proposals depend so heavily on the incorporation of local participation into their schemes. This was an area of considerable poverty, and the minister turned to the people who knew it best, the locals. As Martha Honey relates:

> After a series of community meetings, a team of psychologists and sociologists carried out a detailed house-to-house survey, soliciting opinions on all aspects of tourism. The community gave the green light and chose a special neighbour-hood council to handle all issues relating to the hotel and tourism. 'It's a community hotel, developed and managed by the community,' says Jorge Ramón Cuevas, president of ProNaturaleza, a nongovernmental environmental organisation that works with Las Terrazas.[82]

This may be somewhat different from the way in which participation is viewed by promoters of the tourism industry in capitalist countries, but as the Business

Enterprises for Sustainable Travel (BEST), an organisation with its foundations firmly in the capitalist world, reports:

> local radio and newspapers offered a forum for debating its merits. At each step, local labour was used whenever possible, from building the structures to staffing the hotel and serving as drivers and guides . . . even the highly skilled people brought in to run the Ecological Research Centre became part of the community.[83]

Today about one-third of the hotel's net income is allocated to use in the community and over 85 per cent of Las Terrazas' residents are involved in tourism with 5 per cent continuing to work on reforestation in the reserve. Polly Pattullo describes Las Terrazas as characterising 'low-impact technology, community involvement and a harking back to "solidarity" tourism'.[84]

Whatever one may think about some of the commonly used rhetoric used in the above descriptions of Las Terrazas and of the community's unrepresentativeness of Cuban tourism, the participation element of the project is clearly something that might make the directors of many other supposedly community-based tourism schemes involving cosmetic exercises in participation blush. By all accounts and from all perspectives, it would seem that the Moka Ecolodge would merit the description of *interactive participation* by Pretty's typology. Similar exercises in participation can be found elsewhere in the LAC countries, but they rarely arise from government initiatives.

It is also noteworthy that many of the tourism developments which would earn the descriptor of *interactive participation* or *self-mobilisation* are small in scale. For instance, the examples given earlier in this chapter – Tropic Ecological Adventures, Ecuador (Box 3.1); the solidarity tour to El Salvador (Box 3.4); and Finca Sonador in Costa Rica (Box 3.5) – are all small in scale and might all qualify for the higher grades of participation by Pretty's typology. This is not to say that large-scale developments cannot achieve such levels, but there exist precious few instances of large tourist condominia, large resort developments or large all-inclusives in the subcontinent which have not attracted considerable local opposition, which is not quite the kind of participation implied in the efforts of project designers to involve local people.

It is such large-scale schemes to which we now turn in order to examine whether they offer real benefit to local communities.

Tourism mega-projects and the trickle-down effect – is size important?

On first thought, it might seem a little odd to include a section on tourism mega-projects (by which we mean tourism developments requiring large amounts of capital investment) in a chapter focused principally on the community and local developments. But a theme closely related to both of these foci is poverty reduction through tourism, and a claim frequently made for tourism investments on a grand scale is that they provide for local people employment opportunities not otherwise available to them. This may not always be a good thing in terms of the nature of the work or the rate of pay, but it is undeniable that such employment brings into local communities income that would not otherwise be there. In this way, it is claimed, the income invested in the development (through the employment it provides and from the pockets of the tourists it brings) trickles down to all sectors of the local community or communities affected by the development. It is through this trickle-down mechanism that development spreads and poverty in the locality may be reduced – or so the theory goes.

The trickle-down mechanism has been an important justification for capitalist economic development since the mid-twentieth century when President Truman's inaugural speech to the US Congress triggered the starting gun of the race to development. Since that time we have been repeatedly told that we must be patient for the benefits of capitalist development to trickle down to all strata of society, providing that we pursue the policies which promote the conditions for capital investment. These are the policies outlined in Chapter 2, which in general around the LAC countries have led to a tightening of the public purse strings, a reduction in public spending and employment, the gradual lowering of wages, increasing hardship for the majority and an increasingly wide gap between the haves and the have-nots. In over fifty years of development an increasing number of people in the LAC countries are still waiting for the trickle of wealth to reach them, whilst much of the tourism industry repatriates its profits to head offices in the same countries of the north which supply the tourists. As Evelyne Hong outlined over twenty years ago, 'the real beneficiaries are the rich industrialised tourist-generating countries which control the entire industry'.[85]

The criticism of the trickle-down effect is that it is an illusory carrot held out to the national governments of the LAC countries by the international financial institutions, the transnational corporations and First World governments wanting to extend their investments and influence within the target country. The carrot is then offered on by the national government to its people as an excuse for accepting the overtures of the First World for development that will bring little benefit to the country and which will repatriate its resources and income to the investing country.

Figure 3.3 portrays both the contempt with which the critics of neoliberal capitalist economic development view its proponents and the contempt with which those proponents, the business representatives and investors of the First World, view the populations of the LAC countries. Using a few examples from the world of tourism, it is our aim in this section to examine the reality or illusion of the trickle-down effect. The examples chosen are not all huge developments covering large areas of land as the term 'mega-project' may suggest. But all have required in their construction large amounts of investment capital that are unlikely to be found in the locality of the project. Also, all are of the enclave type of tourist development that are exclusive in the sense that visitors to them need to be relatively wealthy in order to enter, which means that the majority of local people are automatically excluded unless of course they are employed to service the visitors.

Figure 3.3 Trickle-down theory explained

Sandals, the Caribbean

Our first example is that of the eighteen Sandals resorts on the Caribbean islands of Jamaica, St Lucia, Antigua, Bahamas, Cuba and Turks & Caicos. The Sandals resorts are all-inclusives, in which all the needs and desires of the tourists are met within the boundaries of the resort, and for which the tourist pays one fee at the booking stage of the holiday which covers virtually all costs. This effectively means that within the resort guests can wander around without the need to carry money on their person. The Sandals resorts boast that their all-inclusives are more all-inclusive than all the other all-inclusives – or 'ultra all-inclusive'[86] as a Sandals 2005 advertisement calls them. The Sandals website declares:

> Other all-inclusives may advertise 'everything's included,' but they often disclaim that many of their services and amenities are additional. This includes

such features as beer, wine, premium brand alcohol . . . and tipping. Not to mention such sports as waterskiing, scuba diving and golf. Even government taxes and airport transfers. These, among others, when not budgeted for, can double the cost of your so-called 'all-inclusive' holiday. At Sandals, there are no compromises and no surprises. The best of everything is included in one up-front price.[87]

Sandals specialises in catering for couples and arranges four thousand weddings a year and has what it refers to as 'Sandals Resorts' and 'Beaches Resorts', which together can accommodate guests in over 4,000 rooms. Sandals Resorts are for couples only while Beaches Resorts are for everyone with the exception of the Beaches Resort in Cuba which is for adults and over sixteen years only. This seems to imply that it really does not matter which location within the Caribbean you holiday in – they are all the same within the bounds of the resort, more or less complying with what the company calls the 'Sandals concept'. In turn, this implies that the surrounding environment is of no importance except inasmuch as it may be used for enjoyment, sport or adventure. Surrounding communities are of no import except inasmuch as they house the labour used in the resort and unless they offer some cultural sight or feature that may serve as the object of the tourists' fascination or curiosity.

All-inclusives can be seen as environmental bubbles, separated and guarded from the rest of society, whose only interaction with surrounding local communities is through the daily arrival and departure of the service staff. Polly Pattullo reports on an admittedly rather dated 1991 survey[88] which showed that whilst 73 per cent of hotel workers on the island of Jamaica complained of low wages and benefits, at Sandals only 35 per cent made the same complaint.[89] This might suggest a relatively enlightened and progressive labour policy pursued by the company. On the other hand, in the first edition of her book, she also reports that when Sandals took over the La Toc Hotel in St Lucia in 1993, 'many of the 300 or so workers were allegedly either not re-employed with the new company or were offered jobs in lower positions and at lower rates',[90] which might suggest the opposite. And in the second edition of her book she reports that the Sandals chain remains non-unionised, except in Antigua.[91]

Pattullo also reports on a 1994 financial survey of accommodation types in Jamaica[92] which found that 'all-inclusive hotels generate the largest amount of revenue but their impact on the economy is smaller per dollar of revenue than other accommodation subsectors'. It also concluded that 'all-inclusives imported more and employed fewer people per dollar of revenue than other hotels. This information confirms the concern of those who argue that all-inclusives have a smaller trickle-down effect.'[93]

A final point about Sandals is that its owner, Butch Stewart, is Jamaican and the company cannot be accused of repatriating its profits to a head office in the capital of a G8 nation. This does not mean that its profits remain on all the islands on which it operates ready for re-investment there. It is repatriated instead to Jamaica, and, according to its own press release, it provides the state of Jamaica with 10 per cent of its hard currency. This is impressive if correct, but it needs to be emphasised that this does not mean that any of this money finds its way into the local communities most closely affected by their resorts.

Marina Puesta Del Sol, Nicaragua

Our second example is a small mega-project by the standards of a company such as Sandals, but has required considerable financial investment to date. It is the Marina Puesta Del Sol in the Pacific coastal community of Aserradores in the Cosigüina Peninsula of Nicaragua. The Cosigüina Peninsula is an agricultural area characterised by rural poverty surrounded by coastal communities such as Aserradores in which most families have traditionally made their living by artesanal fishing and small-scale shrimp farming. Few of the 143 communities in the area have electricity and outside the main town of El Viejo the water supply is from wells or rivers and the most sophisticated sanitation system is a sump latrine. The village of Aserradores, in which is located the Marina Puesta Del Sol, has a population of approximately five hundred.

The Marina Puesta Del Sol is the project of Roberto Membreño, a millionaire US businessman who was born in Nicaragua. Begun in the year 2001, the marina and hotel complex, when finished, will have around 200 berths (which is small by modern yachting marina standards), and a hotel offering 260 beds, a driving range, a golf course, swimming pool, tennis courts, air strip, helipad, restaurants, shops and many other smaller facilities for guests. In early 2005, the hotel had 33 luxury

Figure 3.4 Marina Puesta Del Sol, Nicaragua
Photo: Martin Mowforth

suites. Figure 3.4 shows the main berthing point of the marina next to the quay, the hotel and swimming pool.

Although the development has other investors, Mexican and North American, it is closely and personally identified with Sr Membreño, a personal friend of Enrique Bolaños, the Nicaraguan president until the end of 2006. The development is rumoured to involve an investment of at least US$100 million on the part of Sr Membreño.

The care and thoroughness with which the construction work is carried out and the few yachts that have so far berthed there suggest that this will be a resort from which even the mildly rich might find themselves excluded on economic grounds. But Membreño has been keen to focus some of the propaganda and publicity about his project on the benefits to the local community. Interviewed on site in May 2003, he not only remarked on the quality of workmanship but also pointed out the spin-offs for the local workmen: 'Look at this. It's superb. My carpenters from Corinto have taught the local workers to do things like this. They have skills now that they could not have dreamed of before.'[94] Moreover, consider this endorsement from two of his yachting guests referring to Membreño and one of his event organisers, Gene Menzie: 'these guys care at least as much about the local people as they do about profits. We've talked to Membreño a number of times, and he's giddy about the project in a large part because of the employment and educational opportunities it is and will afford the locals.'[95]

Membreño recognises that few of the locals will be able to provide the services he requires for his guests in Marina Puesta Del Sol without training. He argues forcefully, however, that not only are the benefits of his seafaring tourism development trickling down to the locals through the construction phase of the project in which some of them are hired as labourers and trainees, but also that some of them can be trained as service providers, attendants and in other posts when the development is fully up and running. He also believes, credibly, that a development such as Marina Puesta Del Sol will give rise to a range of other related businesses in the vicinity. There is little doubt that the potential for this kind of benefit in the community is great, and it is hard to find anyone in Aserradores who does not acknowledge that some local people have benefited and will continue to benefit through employment in the development.

It is not, however, difficult to find local opposition to the marina – the story of Marina Puesta Del Sol is not as clear-cut and as widely beneficial as the developer's publicity would have us believe. The development has necessitated the purchase of a considerable amount of land in Aserradores, most of which is now policed by private guards and protected by barbed wire and 'Private Property' notices. The fishermen of the community have been left with only one narrow access point to the estuary, now crowded with boats; access to the estuary and the sea is prohibited in all but a few points which are not owned by Sr Membreño; most of the land previously owned by the Mario Carrio Chevez fishing co-operative in agreement with the municipality was sold off, in dubious and undemocratic circumstances, to Sr Membreño; the lack of suitable provision of potable water to residents of Aserradores is a major issue for the residents, especially when set beside the luxury provision of a swimming pool and a number of hydro-massage whirlpools; two short stretches of estuarine mangrove vegetation have been destroyed to be replaced by white sand imported from the Pacific Ocean coast of the area; and heavy pressure is being exerted on one particular family to quit their land which forms something of an island surrounded by the Marina Puesta Del Sol.

On the destruction of mangroves either side of the major quay to make way for white sand brought in from the nearby Pacific coast, in 2003 Sr Membreño was fulsome in his praise of the geologist who had assured him that the respective ecosystems would not be adversely affected. It has to be remarked, however, that it would be difficult to find an environmentalist anywhere in the world or a geologist not in Sr Membreño's pay who would agree that such actions would have no effect.

The problem of water supply in the community is probably the one issue that generates the most heated sentiment. For years the community has had problems with its water supply. Water in the wells in the immediate area is too saline to be potable and subterranean water in general in this area is too contaminated with agrochemicals to be of use. In 2003 the supply of water was identified as the highest priority for resolution by over half the population, the next major problem (the lack of electricity) being identified by less than 20 per cent.[96] Francisco José Maliaño Molina, a former leader of the community now retired and tending his herd of goats, explains how in 2002 he sent a letter to Sr Membreño, signed by over 80 residents, asking him to take into account in his development the community's need for potable water. He never received a reply and concludes that: 'he's not interested in the community'.[97] Juan Alberto Chieres Casco, a member of the local administrative committee, relates that 'members of the local committee have discussed the problem [of water suppy] with him [Sr Membreño] and he suggested that he was going to do something about it; but we don't know when'.[98]

Regarding the land ownership in and around the community, various residents report that Sr Membreño's supposed benevolence towards the local community has concealed his aggressive approach to acquiring the land he required for the development. The family of Max Garay in particular has felt threatened by the acquisition of land for the marina and hotel complex. Sr Garay is one of the few Nicaraguans resident in Aserradores who own sizeable parcels of land. He explains:

> The [fishing] co-operative gave up its rights to the Marina Puesta Del Sol and the marina began to develop a hostile attitude towards various residents, including my family. My wife's uncle, for instance, was cruelly evicted from his land of which he was the legal owner . . . He [Sr Membreño] has managed, through buying influence, to turn the illegal into the legal. . . . Bit by bit he has closed down our spaces along the estuary and at his whim they have closed our access to the sea for subsistence . . . The judicial system in Nicaragua is very easy to corrupt.[99]

Max Garay's wife, Tadea, also interjects: 'I wanted to talk directly with him so that we could put all our cards on the table and tell him that I am not against his project. Our country needs progress. The only thing that I am asking is that he respects our rights.'[100] And Allan Bolt, a journalist with *El Nuevo Diario*, one of Nicaragua's leading daily newspapers, confirms this version of events:

> Everybody, including the Garay family, welcomed the new tourism project enthusiastically because it meant work and increasing affluence for all. But it seems that this investor has his own vision of how he wants the countryside to appear and what type of people he wants to see there, such that he has closed the public right of way to the shore (which is unconstitutional, but which the authorities have allowed), he has prohibited his employees from making purchases in Garay's *pulperia* [mini-store], he has tried to throw them off the Island of Aserradores (despite their land titles), and he has been supported in this dirty game by all the powers of the state . . . But if we aren't institutionally prepared, and if our officials do not work for the common good, then the big fish from outside are going to gobble up all the local fish, large, medium and small.[101]

In this case there seems little doubt that some benefits have trickled down and will continue to trickle down to some members of the community of Aserradores and beyond. These appear to have been offset to an uncertain extent by the trickle-down of a number of disbenefits to various residents, and it is an impossible task to quantify the net effect of these different impacts. But putting aside all the verbose and eulogistic publicity material in favour of the development, all the local criticisms against it and all doubts about the very notion of the trickle-down effect, probably the most telling comparison to make is the fact that the amount of money invested in the Marina Puesta Del Sol could have provided a safe potable water supply system for every person in the whole of the Cosigüina Peninsula. Such an investment would have spread the health and security benefits of this amount of money to a huge number of people instead of leisure benefits to a small number of people whose wealth is already great enough to ensure their basic human needs and rights.

Papagayo, Costa Rica

The *Ecodesarrollo Papagayo* (Papagayo Ecodevelopment) on the Pacific coast of Costa Rica is well known and well documented as a tourism megaproject. It was initially conceived on a grand scale as the 'Cancún of Costa Rica' with vacation homes, condominia, shopping centres, golf courses, marinas, hotels, polo grounds and many more facilities covering seventeen beaches and providing twice as many rooms for visitors as were available in the whole of the rest of the country. The development required an investment of US$3 billion and was begun in the early 1990s when the then Minister of Tourism signed an agreement with Grupo Situr, a Mexican resort development company. The government gave the company a long lease, tax exemptions, and large plots of land. It also committed itself to providing the infrastructure necessary for the development.

Partly as a result of a catalogue of complaints and denunciations, the development had ground to a halt by the mid-1990s but was given the green light again by a new government and Minister of Tourism. The green light was for development on a much smaller scale, however, and 'Cancún' has not yet been re-created in Costa Rica. Nevertheless, development continued in the designated area as Jean McNeil explains:

> with hotels with golf courses being the latest development craze – a seemingly mad idea in this semi-arid landscape. Environmentalists fear that the enormous amount of water needed to maintain these golf courses will be abstracted from wetlands, mangroves and other delicate habitats, whilst the locals are worried that their water supplies may be curtailed.[102]

In 1999, the Ecodesarrollo Papagayo company along with the Costa Rican Institute of Tourism (ICT by its Spanish initials) developed a new master plan for the 1,000 hectare Peninsula Papagayo. According to the publicity of the Gaia Group which provides the planning and design services to the project, the new plan was to include two large-scale resort complexes, at least six boutique hotel sites, two or three golf courses, beach clubs, a marina and 900 high-class home sites. The publicity describes how the plan will embrace 'environmental sensitivity', 'ideals of environmental stewardship' and 'sustainable design'.[103]

The Four Seasons Resort Hotel was inaugurated in January 2004 at the same time that local Guanacastecos and environmentalists staged an action to highlight the barring of public access to the peninsula's beaches. Claims and counter-claims were thrown around, but the essence of the argument was caught in the headline of a press release issued by the protesting groups (made up of three local groups and the national Costa Rican Federation for Environmental Conservation [FECON]): 'Four Seasons Takes Over Playa Blanca and Playa Virador: Opens its Doors to Jet-Set Tourism and Closes the Beaches to Tico Tourists.'[104] (*Tico* is a colloquial word for a Costa Rican.)

The details of the case are not vital for the purposes of this book. Suffice it to note that, once again, the trickle-down effect was not sufficient to persuade all local people that the project was wholly beneficial. Despite publicity hyperbole about environmental and cultural sensitivity, the rights of local people take second place to the needs and expectations of foreign tourists and the profits of foreign corporations. As Martha Honey reports, in the words of a Costa Rican critic of the scheme, 'The only thing green about these places is the dollars they are earning'.[105]

Despite its international linkages and dependence on foreign clients, tour operators, airlines and corporations, it is undeniable that the developments at Papagayo have created a large number of service jobs. These undoubtedly contribute to a trickle-down effect in terms of income entering local communities and the rest of the country, but to some extent such benefits are counterbalanced by the loss of the rights to ownership of land, the loss of the right of the majority to use beaches which have been privatised and made exclusive, and the decrease in the availability of water resources to local communities – see Figure 3.5. As Jeff Marshall described the scheme in 1994, 'It is nothing more than a high-profit real estate scheme designed to make a bundle of money for a few Costa Rican insiders and their foreign

Who decides how to allocate water resources?

Figure 3.5 Conflicts over water resources in Guanacaste, Costa Rica
By kind permission of the *Tico Times*

corporate allies'.[106] Martha Honey puts more flesh on this critique and describes how the Papagayo development became one of the spark plugs for the motivation of the Costa Rican environmental movement which in turn exerted pressure on the Costa Rican Institute of Tourism to alter its regulations to permit the development of smaller lodges and cabins with just ten rooms to qualify for incentives and tax breaks. She quotes a 1997 interview with Ottón Solís in which he explains that despite the changes in the regulations, 'To qualify you need a tourism contract, and to get a contract you have to have consultants and present feasibility studies, and all that is possible only for large companies. The big hotels get the tax exemptions, and the little things are left for locals.'[107]

Los Micos Beach and Golf Resort, Honduras[108]

The Honduran Chamber of Tourism's website invites tourists to Tela Bay with the enthusiastic description of:

> A sleepy coastal town that is rapidly becoming a favorite vacation alternative, Tela has grown . . . into the premier beach vacation destination in mainland Honduras. Its proximity to three national parks, to Garífuna villages and its

outstanding beaches guarantee a sustained growth in the tourism industry for years to come.[109]

Tela Bay is the site of the Los Micos tourist development, a multi-million-dollar project funded by the Inter-American Development Bank (IDB), the Italian government, Honduran business and at least one European bank. The Los Micos development is slated to include many of the trappings of a large, mainstream holiday resort, including hotels run by Hilton and Hyatt, malls, an equestrian centre and golf courses, as well as a cruise ship dock in the town of Tela.[110] International investors and the Honduran government claim that the development will bring significant benefits to the area, including jobs and infrastructure, and funding for the site is included in $35 million of IDB money specifically earmarked for a 'national sustainable tourism programme' for Honduras.[111]

There has been considerable resistance to the development, however, mainly from the Garífuna community. The Garífuna have lived in the area for over 200 years, and are descended from the Arawak peoples of the Caribbean islands and escaped Africans who were trafficked into slavery. They have maintained communal land ownership structures, and their land title in the Tela Bay area was established in 1992.[112] Previous developments in the area have seen the Garífuna come into conflict with the national government and municipal councils over land rights, including a luxury tourism development started in 1994, which now lies empty.[113] In 1997, the national government is alleged to have 'conveniently "lost"' documents relating to Garífuna title to the land.[114]

The challenges to Garífuna land rights are also part of a wider pattern, supported by international institutions such as the World Bank and WTO/OMC, which is seeking to shift communal forms of land rights, as practised by many indigenous peoples in Central America, to systems of individual rights which are easier for international economic actors to deal with.[115] Such communal land rights have been key to many examples of indigenous resistance to developments such as mining, tourism and bioprospecting in Mexico and Central American countries.[116]

As well as institutional attacks on their land title, the Garífuna of Tela Bay have also suffered direct human rights abuses aimed at forcing them to relinquish their rights to the Los Micos development land. According to a report submitted to the UN Human Rights Committee in October 2006 by US NGO Human Rights First, which has supported the Garífuna claims, the major incidents have included the following.

*   The shooting of Gregoria Flores Martínez, the General Co-ordinator of OFRANEH, the main Garífuna community organisation fighting the Los Micos development. Ms Flores was shot after a series of warnings regarding her campaigns for Garífuna land rights and while collecting testimonies regarding the alleged false imprisonment of another community leader.[117] The Inter-American Court of Human Rights issued a resolution acknowledging the precarious situation for Garífuna activists and asked for protective measures for Ms Flores and her family, which were not implemented.[118]

- The alleged false imprisonment of Alfredo López Alvarez, a leading member of several Garífuna rights organisations, who was arrested in 1997 on drugs charges, found guilty in 2000, exonerated in 2001 and January 2003, but not released until August 2003. In February 2006 the Inter-American Court of Human Rights condemned the Honduran authorities for their detention of Mr Alvarez and ordered them to pay reparations, which have not been forthcoming.[119]
- The burning of the house of Wilfredo Guerrero, the President of the Committee to Defend the Lands of San Juan, the site of the Los Micos complex. Although no one was hurt in the fire, documents vital to the Garífuna case were destroyed.[120]
- Threats to the life of Jessica García and her children. Ms García, a Garífuna community leader, was approached at home in June 2006 by a man who offered her money to sign a document surrendering Garífuna land rights to the development company PROMOTUR. When she refused, the man put a gun to her head to force her to sign, and threatened her life and those of her children if she publicised the document's existence.[121] The document, a copy of which was obtained by a US human rights group, is said to hand the disputed territory over to PROMOTUR, guarantee that the Garífuna would abandon legal actions or complaints, and that PROMOTUR would have the right to evict and relocate Garífuna communities. The document is said to have been co-signed by PROMOTUR owner Jaime Rosenthal Oliva.[122]

Activists involved in the land rights campaign at Tela Bay have also alleged corruption against the Honduran government and local authorities. Yani Rosenthal Hidalgo, a recently appointed minister to the Honduran government, is the son of the owner of PROMOTUR and a shareholder in the Los Micos project.[123] Alfredo López's colleague Jesús Alvarez also accused the municipality of Tela of embezzlement in relation to earlier tourist developments. He died following the second of two murder attempts.[124]

As well as the threats to their direct land rights, the Garífuna have also questioned the environmental impacts of the development and rejected an Environmental Impact Assessment which projected benefits to the area. The Garífuna claim that the Los Micos site will increase pesticide use and eutrophication of lagoons owing to fertilizers used on golf courses, as well as put pressure on the area's water resources.[125]

Much has been made in publicity material for Los Micos of its situation within the boundaries of the Jeanette Kawas National Park, and of the potential for ecotourism on the site. The Garífuna claim that the declaration of national parks and other environmental protection areas has been used as a means of violating their territorial rights and of preventing them from carrying out basic subsistence activities. They criticise actors such as the WWF and Smithsonian Institute, which declared the nearby Cayos Cochinos islands a national park in 1993. The Garífuna state that since 1993 they have been prevented from subsistence fishing, including by interference from a naval base, while large-scale commercial fishing has been allowed to continue.[126] They also allege that the creation of ecological reserves is a deliberate method of dissociating them from control of their land, 'so that later the protected areas can be raffled off among the same old sorcerers as always'.[127]

Certainly, the claim that a development of four- and five-star hotels with malls and golf courses can be a source of 'sustainable' tourism or ecotourism demonstrates the lack of a clear definition of such projects. The conflict between the Garífuna and international nature conservation organisations represents a struggle for power and is one of a growing number of such clashes between indigenous subsistence and concepts of the purity of nature. In late 2006, the Honduran government announced that two of the contractor companies involved in Los Micos had been removed, and the IDB announced $1.37 million in funding for capacity building in 'ethnic minority' (note: not indigenous or first nations) tourist enterprises in Honduras, through 'training, assistance and micro-credit'.[128] It remains to be seen whether such concessions are matched with improved economic and human rights conditions.

Concluding remarks

The incidence of poverty around the LAC countries has grown into something of a problem, mostly of course for the poor themselves but also for the promoters, cheerleaders and apologists for the prevailing model of capitalist economic development. If poverty is growing in both absolute and relative terms, then it signifies that the prevailing economic order is not working as it should. Such a situation would not present a problem if it were not noticed, but it is, increasingly so – and not just by the growing army of the poor, but also by the liberal middle classes of the First World whose conscience tells them that their wealth should not be dependent upon the exploitation of the resources and labour of unseen poor people on the other side of the world.

Enter stage right into the world of development, through hope and propaganda, the tourism industry – responsible, sustainable, alternative, and community-based – a new, young industry, apparently clean without smoking chimneys, that brings enjoyment to its clients and income to its local employees and service providers. But all the hope and money invested in the industry still has not managed to alter the incidence of poverty, at least not in the right direction. So, enter stage left the world of pro-poor tourism, designed to stimulate poverty reduction and even poverty elimination. There exist questions of whether such initiatives only confirm the prevailing structures of power and privilege or challenge them, but it is generally accepted that, although the benefits of pro-poor tourism may be insignificant in national and international terms, in local terms they can be very significant.

The growth of community-based tourism initiatives, which led into pro-poor tourism, ran aground on the spin of the politicians who described it as the way forward for their national industries but who at the same time gave incentives only to the large-scale, transnational tourism enterprises. The activities of the latter were aimed at reducing to the minimum the economic benefits that might be left within the local community in order, of course, to maximise the profits to be repatriated to head offices and shareholders in the rich countries. Genuine efforts to involve local communities run into difficulties through their assumption that communities exist as homogeneous entities in which everyone thinks and acts alike, when in reality differences within communities lead to the formation of elite groups who exercise power to ensure the continuation of their favoured position. Normally with the best of intentions, NGOs and INGOs have promoted community-based tourism initiatives in the belief that the benefits will reach everyone in the community. But while control of the initiative remains in the hands of the outsiders – the NGOs if not the foreign corporation – the benefits will rarely be well distributed around the local community.

Throughout this recent period of well-intentioned efforts to ensure that the financial benefits of tourism are felt by local people and to reduce poverty through tourism, one word has risen to prominence: 'participation'. It has become something of a buzzword such that no report or proposal is complete without it. But participation has different significances to different people, as Jules Pretty's typology of participation illustrates; and it is important for analysts and students of tourism developments (not to mention tourists themselves) to examine the notion of participation with this understanding in mind rather than to accept the use of participation in any development project without question.

The common perception of tourism mega-projects is that they are too grand and too tied to the international sector to concern themselves with the local communities whose space they occupy. This may or may not be so, but regardless of the attitude of the mega-project management, it is also widely recognised that such tourism developments offer people in local communities a range of employment opportunities not otherwise available to them. This may not always be a good thing in terms of the nature of the work or the rate of pay, but it is undeniable that such employment brings into local communities income that would not otherwise be there. Notwithstanding the leakage of tourism income from tourism mega-projects back to the base country of the corporation which runs the facility, the income earned, however meagre, then has a multiplier effect within the community. We examined a number of tourism mega-projects and found that whilst the existence of the trickle-down effect was confirmed, it is often minimal in significance and is counterbalanced by disbenefits brought into local communities by the project. These disbenefits may include factors that are crucial to the everyday life of local people such as land tenure and resource use, and the removal of these rights is often less noticeable (but is no less important) than the positive effect of the income that trickles into the community.

Notes

1 GDP is the total value of the output of goods and services produced by an economy, by both residents and non-residents, regardless of the allocation to domestic and foreign claims. The gross national product (GNP) is the total domestic and foreign value added claimed by residents and therefore equals the GDP + net income from abroad (which is the income residents receive from abroad for services (labour and capital) less similar payments made to non-residents who contribute to the domestic economy).

2 Tim Griffiths (1996) 'El Salvador: Economic Round-up', *Mesoamerica* 15, 2, February, Institute for Central American Studies, San José, Costa Rica.

3 United Nations Economic Commission for Latin America and the Caribbean (ECLAC) (2002) 'Poverty Increased in Lost Half-decade', Comment by José Antonio Ocampo, ECLAC Executive Secretary, ECLAC Notes No. 25, November 2002.
4 World Bank (annual) 'World Development Report', New York: Oxford University Press.
5 New Internationalist (2003) 'The Liberation of Latin America', No. 356, May, Oxford: New Internationalist Publications, p. 18.
6 Vandana Shiva (2005) 'Hacer que la pobreza sea historia, y la historia de la pobreza', Znet on www.zmag.org.
7 The Office of the UN High Commissioner for Human Rights: www.unhchr.ch/development/poverty-01.html.
8 UNDP (annual) *Human Development Report*, Oxford: Oxford University Press.
9 UNDP (1996) *Human Development Report*, Oxford: Oxford University Press, p. 67.
10 *Ibid*. p. 67.
11 Clifton Holland (2005) 'The OAS at the Crossroads', *Mesoamerica* 24, 8, August, San José, Costa Rica.
12 UNDP (2003) *Human Development Report*, Oxford: Oxford University Press, p. 17.
13 Oliver Marshall (2004) 'Introducing South America', *Rough News* 23, Summer, London: Rough Guides Ltd.
14 José Antonio Ocampo (2002).
15 Bridget Wooding and Richard Moseley-Williams (2004) 'Worlds Apart', *Interact*, Spring, London: Catholic Institute for International Relations.
16 Deloitte and Touche, International Institute for Environment and Development (IIED) and Overseas Development Institute (ODI) (1999) 'Sustainable Tourism and Poverty Elimination Study: A Report to the Department for International Development', Unpublished Report, Department for International Development, London.
17 Caroline Ashley, Dilys Roe and Harold Goodwin (2001) *Pro-poor Tourism Strategies: Making Tourism Work for the Poor: A Review of Experience*, London: Overseas Development Institute.
18 Deloitte and Touche *et al*. (1999), p. 56.
19 *Ibid*., p. 7.
20 *Ibid*., p. 8.
21 *Ibid*., p. 22.
22 Ashley, C. *et al*. (2001), p. 28.
23 *Ibid*., p. viii.
24 Deloitte and Touche *et al*. (1999).
25 Ashley, C. *et al*. (2001). Case study research covered South Africa, Nepal, Namibia, Uganda, Ecuador and St Lucia.
26 *Ibid*., p. 11.
27 The Pro-Poor Tourism website has been created by the Pro-Poor Tourism Partnership, a collaborative research initiative between the International Centre for Responsible Tourism, the International Institute for Environment and Development and the Overseas Development Institute.
28 Xavier Cattarinich (2005) 'Pro-poor Tourism Initiatives in Developing Countries: Analysis of Secondary Case Studies', Pro-Poor Tourism Partnership website: www.propoortourism.org.uk.
29 Ashley, C. *et al*. (2001) p. 28.
30 Deloitte and Touche *et al*. (1999), p. 9.
31 *Ibid*., p. 26.
32 Cox, A. (1999) 'DAC Scoping Study of Donor Poverty Reduction Policies and Practices', Synthesis Report, ODI and ARID for the DAC Informal Network on Poverty Reduction.
33 Ashley, C. *et al*. (2001), p. 39.

34 Tourism Concern's *In Focus* journal regularly features clashes between local communities and tourism but in this regard its Spring 2002 issue (number 42) is particularly noteworthy.

35 Flynn, M. (1996) 'Report on Guatemala', *Mesoamerica* 15, 8: 3–4.

36 Tourism Concern (2002) 'Ecotourism Evictions', *In Focus*, 42, London: Tourism Concern.

37 Polly Pattullo and Orely Minelli for Tourism Concern (2006) *The Ethical Travel Guide*, London: Earthscan. This is the third edition of *The Community Tourism Guide* also published by Earthscan in 2000 for Tourism Concern and written by Mark Mann who also wrote the second edition entitled *The Good Alternative Travel Guide* in 2002 and again published by Earthscan.

38 Mark Mann for Tourism Concern (2002) *The Good Alternative Travel Guide*, London: Earthscan, p. 19.

39 *Ibid.*, p. 19.

40 Deborah Dwek (2004) 'Favela Tourism: Innocent Fascination or Inevitable Exploitation?' MA Dissertation, Institute of Latin American Studies, University of London, p. 27.

41 *Ibid.*, p. 30.

42 Mann (2002) p. 20.

43 Judy Bennett (1997) 'San Blas: The Role of Control and Community Participation in Sustainable Tourism Development', unpublished MA dissertation, University of North London.

44 Quoted by David Atkinson (2004) 'The Che Revolution', *Developments* 27, London: Department For International Development.

45 Quoted by David Atkinson *ibid*.

46 *Ibid.*

47 Anita Pleumarom (2004) 'Bolivia – Che Guevara's Death Place to Become a Tourist Draw', Clearinghouse of the Tourism Investigation and Monitoring Team, Bangkok, Thailand, 30 August.

48 James Fair (1996) 'Can Ecotourism Save Ecuador's Threatened Cloud Forests?' in Helen Collinson (ed.) *Green Guerrillas: Environmental Conflicts and Initiatives in Latin America and the Caribbean*, London: Latin America Bureau.

49 *Ibid.*, p. 120.

50 Anita Pleumarom (2003) TIM-Team Clearinghouse bulletin, 12 August 2003.

51 Rosie Mayston (2004) email communication, 17 June 2004.

52 David Barkin (2000) 'The Economic Impacts of Ecotourism: Conflicts and Solutions in Highland Mexico', in P. M. Godde, M. F. Price and F. M. Zimmerman, *Tourism and Development in Mountain Regions*, New York: CAB International.

53 James Fair (1996), p. 120.

54 Jules Pretty (1995) 'The Many Interpretations of Participation', *In Focus* 16, London: Tourism Concern, p. 4.

55 World Bank (1996) *The World Bank Participation Sourcebook: Environmentally Sustainable Development*, Washington DC: World Bank.

56 Henkel, H. and Stirrat, R. (2001) 'Participation as Spiritual Duty: Empowerment as Secular Subjection', in B. Cooke and U. Kothari (eds) *Participation: The New Tyranny?* London: Zed Books, p. 168.

57 Cleaver, F. (2001) 'Institutions, Agency and the Limitations of Participatory Approaches to Development', in B. Cooke and U. Kothari (eds) *Participation: The New Tyranny?* London: Zed Books.

58 Bill Cooke and Uma Kothari (2001) *Participation: The New Tyranny?* London: Zed Books. Desai, V. (1995) *Community Participation and Slum Housing: A Study of Bombay*, London: Sage. Rahnema, M. (1992) 'Participation', in W. Sachs (ed.) *The Development Dictionary: A Guide to Knowledge as Power*, London: Zed Books.

59   Henkel and Stirrat (2001), p. 172.
60   Brandon, K. (1993) 'Basic Steps toward Encouraging Local Participation in Nature Tourism Projects', in K. Lindberg and D. Hawkins (eds) *Ecotourism: A Guide for Planners and Managers*, North Bennington, VT: Ecotourism Society, p. 136.
61   Drake, S. (1991) 'Local Participation in Ecotourism Projects', in T. Whelan (ed.) *Nature Tourism*, Washington DC: Island Press, p. 132.
62   Epler Wood, M. (1991) 'Global Solutions: An Ecotourism Society', in T. Whelan (ed.) *Nature Tourism*, Washington DC: Island Press, p. 204.
63   Cleaver, F. (1999) 'Paradoxes of Participation: Questioning Participatory Approaches to Development', *Journal of International Development* 4, 11; and (2001) 'Institutions, Agency and the Limitations of Participatory Approaches to Development', in B. Cooke and U. Kothari (eds) *Participation: The New Tyranny?* London: Zed Books.
64   Cleaver (2001), p. 37.
65   Michel Foucault (1980) *Power and Knowledge*, Hemel Hempstead: Harvester Wheatsheaf.
66   Uma Kothari (2001) 'Power, Knowledge and Social Control in Participatory Development', in B. Cooke and U. Kothari (eds) *Participation: The New Tyranny?* London: Zed Books, p. 148.
67   Mosse, D. (2001) 'People's Knowledge, Participation and Patronage: Operations and Representations in Rural Development', in B. Cooke and U. Kothari (eds) *Participation: The New Tyranny?* London: Zed Books.
68   Bill Cooke (2001) 'The Social Psychological Limits of Participation', in B. Cooke and U. Kothari (eds) *Participation: The New Tyranny?* London: Zed Books, p. 111.
69   Mosse, D. (2001), p. 24.
70   Taylor, H. (2001) 'Insights into Participation from Critical Management and Labour Process Perspectives', in B. Cooke and U. Kothari (eds) *Participation: The New Tyranny?* London: Zed Books.
71   Cleaver (2001), p. 44.
72   Massey, D. (1991) 'A Global Sense of Place', *Marxism Today*, 24–9 June; (1993) 'Questions of Locality', *Geography* 78, 2: 142–9; (1995) *Spatial Divisions of Labour*, Second edition, London: Macmillan; Hall, S. (1995) 'New Cultures for Old', in D. Massey and P. Jess (eds) *A Place in the World? Places, Cultures and Globalisation*, Milton Keynes: Open University Press.
73   Taylor, H. (2001), p. 137.
74   Brandon, K. (1993), p. 135.
75   Johnston, B. (1990) 'Introduction: Breaking out of the Tourist Trap', *Cultural Survival Quarterly* 14, 1: 2–5.
76   Wells, M. and Brandon, K. (1992) *People and Parks: Linking Protected Area Management with Local Communities*, Washington DC: The World Bank, The World Wildlife Fund and US Agency for International Development.
77   West, P. and Brechin, S. (eds) (1991) *Resident Peoples and National Parks*, Tucson: University of Arizona Press.
78   Jules Pretty (1995) 'The Many Interpretations of Participation', *In Focus* 16: 4–5, London: Tourism Concern. Adapted by Pretty, J. and Hine, R. (1999) *Participatory Appraisal for Community Assessment: Principles and Methods*, Centre for Environment and Society, University of Essex.
79   Mundo Maya Travel Guide (undated), p. 4.
80   Belize Tourism Industry Association (1992) *Tourism Link*, April, Belize City: BTIA, p. 4.
81   Interviewed by Martha Honey and reported in Martha Honey (1999) *Ecotourism and Sustainable Development: Who Owns Paradise?* Washington DC: Island Press, Chapter 6.
82   Honey (1999), p. 183.

83 Business Enterprises for Sustainable Travel (BEST) (2004) 'Community Profile: Las Terrazas Complejo Turístico', BEST website: www.sustainabletravel.org.

84 Polly Pattullo (2005) *Last Resorts: The Cost of Tourism in the Caribbean*, Second edition, London: Latin America Bureau, p. 255.

85 Evelyne Hong (1985) See the Third World While It Lasts, Penang: Consumers' Association of Penang, p. 21.

86 Sandals (25 January 2005) Personal communication.

87 Sandals website (2005) www.sandals.co.uk/general/included.html.

88 Carl Stone (1991) 'A Socio-Economic Study of the Tourism Industry in Jamaica', *Caribbean Affairs*, Trinidad 4, 1.

89 Polly Pattullo (2005 – second edition), p. 69.

90 Polly Pattullo (1996 – first edition) *Last Resorts: The Cost of Tourism in the Caribbean*, London: Cassell, p. 57.

91 Polly Pattullo (2005 – second edition).

92 Organisation of American States (OAS) (1994) *Economic Analysis of Tourism in Jamaica*, Washington DC: OAS.

93 Polly Pattullo (2005 – second edition), p. 97.

94 Roberto Membreño (2003) In discussion with Martin Mowforth and others, Aserradores, 25 April.

95 Changes in Latitudes website (2004) Yachting report, January 2004, www.latitude38. com.

96 SELVA (2003) 'Información Linea Base para el Diseño de Proyectos de Desarrollo de la Comunidad de Aserradores', El Viejo, Nicaragua: SELVA, November.

97 Francisco José Maliaño Molina interviewed by Martin Mowforth, September 2004.

98 Juan Alberto Chieres Casco interviewed by Martin Mowforth, September 2004.

99 Max Garay interviewed by Martin Mowforth, September 2004.

100 Tadea Garay interviewed by Martin Mowforth, September 2004.

101 Allan Bolt (2002) 'Turismo, soberanía y desarrollo', *El Nuevo Diario*, 19 December, Managua, Nicaragua.

102 Jean McNeil (1999) *Costa Rica: The Rough Guide*, London: Rough Guides, p. 232.

103 Gaia Group website (2005) 'Peninsula Papagayo – Guanacaste, Costa Rica: Four Seasons Hotel', www.gaiagroup.com/papagayo.html; Bryan Carlson (2004) 'Peninsula Papagayo, Costa Rica: Master Plan for World Class Destination Resort'; Cámara Costarricense de la Construcción website (2005) 'Revista Construcción', www. construccion.co.cr, 4 February 2005.

104 Costa Rican Federation for Environmental Conservation (FECON) website (2004) 'Four Seasons se Adueña de Playa Blanca y Playa Virador: Abre sus puertas al turismo de jet-set y cierra las playas al turista tico', 19 January 2004, www.feconcr. org; FECON website (2003) 'La Ilegalidad Manda en Papagayo: En presencia del Ministro de Turismo, Four Seasons cierra ingreso playas', 9 December 2003, www.feconcr.org; Gerardo Zamora (2004) '¿Playas de quién?', *Teletica: 7 Días*, 4 February 2005, www.teletica.com.

105 Martha Honey (2002) 'Paradise lost? With Ecotourism Catching on, There's Hope the Travel Industry Will Clean up Its Act. Clean industry or Culture Killer?', *Boston Globe*, 20 January 2002.

106 Jeff Marshall (1994) 'Papagayo Isn't "Ecotourism"', *Tico Times*, San José, Costa Rica, 8 April, pp. 2 and 36.

107 Martha Honey (1999) *Ecotourism and Sustainable Development: Who Owns Paradise?* Washington DC: Island Press, p. 137.

108 This section on the Los Micos Beach and Golf Resort, Honduras, was written by Sarah Irving – www.sarahirving.net.

109 Honduran Chamber of Commerce website: www.telahonduras.com (accessed November 2006).

110 Honduran Institute of Tourism press release 1 June 2004, 'Italian Government Reiterates Its Support for the Country's First Comprehensive Tourism Complex at Tela Bay'.
111 Inter-American Development Bank press release 4 May 2005, 'IDB Approves $35 Million Loan to Honduras for Sustainable Tourism'.
112 Rights Action (2005) 'The Tourist Industry and Repression in Honduras', Rights Action document 31 August 2005, sourced from www.upsidedownworld.org/main/content/view/66/46/.
113 Honduran Institute of Tourism press release 1 June 2004, 'Italian Government Reiterates Its Support for the Country's First Comprehensive Tourism Complex at Tela Bay'.
114 Sandra Cuffe (2006) 'Nature Conservation or Territorial Control and Profits?', 7 February, www.upsidedownworld.org/main/content/view/194/46.
115 Rights Action (2005) 'The Tourist Industry and Repression in Honduras', Rights Action document 31 August 2005, sourced from www.upsidedownworld.org/main/content/view/66/46/.
116 See for example Duncan Green (1995) *Silent Revolution: The Rise of Market Economics in Latin America*, London: Cassell.
117 Human Rights First (2006) Report to the Human Rights Committee on its consideration of the Initial Report by the Government of Honduras under the International Covenant on Civil & Political Rights, 88th Session, 16–17 October 2006.
118 Human Rights First (2006) 'Garífuna Community Leader in Honduras Threatened with Death', document dated 6 July 2006, www.humanrightsfirst.org/defenders/hrd_women/alert070606_garifuna.asp.
119 Human Rights First (2006) Report to the Human Rights Committee on its consideration of the Initial Report by the Government of Honduras under the International Covenant on Civil & Political Rights, 88th Session, 16–17 October 2006.
120 *Ibid.*
121 *Ibid.*
122 Human Rights First (2006) 'Garífuna Community Leader in Honduras Threatened with Death', document dated 6 July 2006, www.humanrightsfirst.org/defenders/hrd_women/alert070606_garifuna.asp.
123 Sandra Cuffe (2006).
124 Rights Action (2005) 'The Tourist Industry and Repression in Honduras', Rights Action document 31 August 2005, sourced from www.upsidedownworld.org/main/content/view/66/46/.
125 Sandra Cuffe (2006).
126 Rights Action (2006) 'Garífuna Communities Continue to Struggle for Territory', Rights Action document dated 17 October 2006, sourced from www.globalexchange.org/4280.html.
127 *Ibid.*
128 Inter-American Development Bank (2006) 'IDB Fund to Promote Networks of Ethnic Enterprises in Honduran Tourism Industry', IDB press release 9 November 2006.

4 Tourism and the environment

Eco by name, eco by nature?

While this book is primarily concerned with exploration of the often problematic relationships between tourism and society in the LAC countries, it is important to acknowledge that there is much critical concern about the environmental impacts of tourism. It is clear that the boundaries between 'environment' and 'society' are ragged and porous, as is very evident in most interpretations of sustainable tourism and ecotourism, which invariably focus on the challenges tourism poses for both the natural environment and host communities. Given the complexity and diversity of the topic, there is no pretence that this section can do more than explore a number of pertinent dimensions of the debate on tourism and the environment.

The chapter opens by considering the diverse, uncertain constitution of the natural environment before offering outlines of tourism's principal environmental impacts, sustainability and tourism, and ecotourism, which has become such a ubiquitous but contested feature of the tourism landscape in the LAC countries. In recognition of widespread scepticism about the way much of the region's tourism industry has evidently responded to concerns about environmental pressures, this chapter also briefly questions the credentials of ecotourism and associated 'greening' devices such as ecolabelling and certification. There is also consideration of other measures that might be used to confront tourism's environmental impacts, including ecological footprinting, carbon budgets, emissions trading schemes and the rather long-established options for nature conservation in protected areas. The chapter concludes by posing some questions that might disturb the 'new orthodoxy' of criticism that casts mass international tourism in the role of principal villain in the environment and tourism discourse in the LAC countries.

The diversity of 'tourism and the environment'

Three related claims feature strongly in critical discussions of tourism and the environment. First, the success of tourism is strongly dependent on the so-called 'natural environment'. Second, tourism imposes negative impacts on this environmental resource base, to the extent that the well-being of not only human communities and other organisms but also the tourist industry itself may be threatened. Finally, widening awareness of the environmental challenge has made the search for more sustainable tourism practices a central focus for tourism policymakers and managers, albeit with widely differing levels of commitment.

However, these three apparently straightforward claims are merely pointers to a highly complex and contested reality, not least in the context of a large and remarkably varied region like Latin America and the Caribbean (the LAC countries). A basic point here is that the 'environment' with which LAC tourism engages is highly diverse. Also, the boundaries and definition of what is 'natural' in the environment can be problematic. While it is difficult to challenge the 'natural' status of the volcanoes of Middle America and the Andes, the rural landscapes encountered by tourists in highland Guatemala, the Peruvian Altiplano and in many other places have been intensely modified by human action.

The picture is further complicated by the growing diversity of tourism systems and products in the LAC countries. As with the natural environment, so also tourism is far from a uniform process. Any discussion of tourism–environment relationships must therefore confront a bewildering selection of tourism destinations, products, visitors, facilities and activities that are constantly evolving, rather than static. The issues, problems and solutions involved will therefore be highly contingent upon particular combinations of environmental factors, locations, forms of tourism and types of tourist, not to mention differing perceptions and values. Thus the 89 hotels plus attendant infrastructure that have transformed a remote 14 km sandy island on Mexico's Caribbean coast into the mega-resort of Cancún might be regarded by the environmentalist as an appalling barbarism, whereas, for many of its visitors, the wonderfully tame and accessible white sand and turquoise seas are convincingly close to the paradise promised by their tour operator's brochure. At a totally different scale, a well-trampled footpath to a popular river bathing spot might be regarded as evidence of undesirable visitor pressure – or as a reassuring thread through an otherwise hostile wilderness.

The natural environment

What then is the 'natural environment' in the context of tourism? For the LAC countries, the popular imagination – vigorously stimulated by colourful publicity materials and media images – would probably start with exotic ingredients like rainforests and 'jungles', volcanoes, canyons, coral reefs, beaches (white sand, of course), jaguars, dolphins and brilliantly coloured parrots. While a complete catalogue is neither possible nor necessary, it is worth considering some of the LAC countries' physical ingredients simply to confirm how the subcontinent's environmental resources are central to its appeal as a tourist destination. The basic list that follows should be accompanied by questions such as: what is the role and significance of these features for tourism? Are they directly experienced by tourists, and if so, how? Who benefits and who is excluded? Or are they essentially an 'imagined' part of the holiday trip (e.g., jaguars in Central America), or alternatively, operate only in the background (e.g., as do many ecological processes)? What are the impacts of tourism on different natural systems and their components? How vulnerable are they to direct or indirect pressure and disturbance from tourism?

- *Climate*: the promise of sun, warmth and blue skies is the baseline attraction in coastal zones. However, seasonality is also a fundamental environmental influence on tourism; the 'when to go' section is a standard item in any guide book. Not surprisingly the Caribbean is somewhat more alluring in the midst of the North American winter than during its hurricane season, while the airport and Antarctic cruise liner jetties of Ushuaia on Tierra del Fuego will see rather more visitors in the southern midsummer than in July or August, as 'winter and its snowy landscapes change Ushuaia's physiognomy radically'[1] despite claims that 'it is possible to enjoy all year long this city's magic'.

- *High-profile, dramatic physical features*: the Iguaçu Falls, the near-vertical Torres del Paine mountains in southern Chile, volcanoes such as Arenal in Costa Rica, Lake Titicaca, Mexico's Copper Canyon, Rio's Sugar Loaf mountain, the Uyuni salt lake in Bolivia or the Perito Moreno glacier in Argentina ('a 60m-high river of rising, toppling and exploding ice'[2]). Such international landmarks are often the principal or at least a highly significant motive for tourist trips, either as 'direct' destinations or dramatic background landscape features. Within destination zones, individual natural features are promoted to tourists – specific 'deserted beaches', lagoons and lakes, such as the *cenotes* in the limestone of Quintana Roo, Mexico, waterfalls, pools, waterways (as in the Paraná delta in Argentina, frequented largely by domestic tourists, and the canals of Tortuguero, Costa Rica, frequented largely by foreign tourists).

- *More generalised landscapes*: typically perceived and projected as 'exotic' and essentially natural, 'pristine' and 'wild' – for example 'jungle'/'rainforest' in Central America and the Amazon Basin, or 'desert', as in parts of Mexico, Peru and Bolivia. Some tourist circuits are based on visits to a series of spectacular landforms, as represented by a tour offered by Latin Trails:

  > from La Paz to . . . Uyuni. Visit the village of Alota and the Valley of Toads with its rock formations. A visit to the Sol de Mañana and Apacheta, a volcanic zone with rich deposits of minerals and salts, boiling mud and small geysers. Continue to the Green Lagoon for a view of the Llicancahur Volcano . . . to the surroundings of the Red Lagoon, famous for its James flamingos and if you are lucky you can observe the change of color of the Lagoon.[3]

- *Coastal landscapes*: especially beaches of white sand, ideally accompanied by palms and turquoise sea, as promoted in much of the Caribbean. Some of these locations are well-established resorts serving international markets such as Montego Bay, Jamaica, Varadero Beach, Cuba, and Acapulco, Cancún and Zihuatenejo in Mexico. Other seaside resorts are much more dependent on intra-LAC markets (such as Mar del Plata, Punta del Este and resorts of north-east Brazil). Another high-profile and often problematic set of coastal landscapes are the coral reefs found off the Caribbean shores of Mexico, Belize and Honduras, frequently described as the second largest barrier reef in the world.

- *Wildlife*: high value is put on the prospect of seeing the more emblematic species in the wild – the promise of the jaguar in Central America, whales and dolphins (Mexico, Panama, Argentina), or Monarch butterflies in the fir forests of Michoacan, Mexico. An extreme case is tourism in the Galapágos Islands, where the unique fauna are by far the major attraction.

The compulsive attraction of exotic species is suggested in this enthusiastic description by the Macaw Landing Foundation, an Oregon-based NGO, of recently 'discovered' salt-lick sites in the vicinity of the Urubamba River in the Peruvian Amazon:

> Tropical Nature Travel has opened the ultimate premiere parrot-viewing site in all of Latin America! Paititi . . . the latest discovery of Charlie Munn, this clay lick is located just a short boat ride on the Urubamba River from Tropical Nature Travel's newest lodge, Timpia. As if the several hundred Blue and Golds at this clay lick were not enough of a spectacle, another clay lick at Kimaruari . . . is an awesome display of Green-Winged, Scarlet, and Blue and Gold macaws. Then when you think it can't get any better, you find yourself heading up the Urubamba to the Pongo de Mainique Canyon to view the Military Macaws nesting in the cliffs next to a magnificent falls. From here you will see fly-overs by Blue-Headed and Severe Macaws. NOW, it's the best, right? WRONG, there is yet another clay lick on the Sebeti River, where you can watch Blue and Gold, Green-Winged, Scarlet and Severe Macaws flying over the rain forest on their way into and from the clay lick.[4]

In contrast, certain other rainforest organisms are viewed as wholly hostile:

> Of all the critters in the rain forest, ['chiggers'] deserve concern and attention. They are red and so small you can hardly see them. When they bite they leave a red welt that can evolve into a blister in some people. They seem to love elastic, and bite mostly around the ankles and the waist. Several different ointments were tried, including a local remedy.[5]

In many cases, isolation of the environment as the source of attraction for tourism is problematic. The object of the tourists' gaze is frequently a combination of the physical environment and human activity, past or present. Many pre-Colombian monuments gain part of their compulsive allure from their setting; examples include Machu Picchu in Peru, perched on a steep-sided saddle in the Andes, or Maya cities in Mexico and Central America still enclosed by apparently intact tropical forests. Many towns and cities stand out for their impressive sites, ranging from former colonial mining towns like Taxco in Mexico to the splendours of Rio de Janeiro. In other cases, the attraction comes from human engagement with the environment, often invoking a sense of 'conquering' the barriers presented by 'Nature'. This may be achieved in relative comfort, as on a journey on the Chihuahua–Pacífico railway in Mexico, which threads the Sierra Madre via 86 tunnels and 39 bridges. Or the conquest may demand a more vigorous input from the tourist, for instance in the case of whitewater rafting, as per this breathless account of a descent of the Futuleufu river in Chile:

> our river guide and president of Earth River Expeditions, is directing us around a rock the size of my garage . . . I lean out over the bow of our raft, laughing like some half-mad figurehead, and drive my paddle down into the silver froth. The current under the boat is pulling hard to the middle of the river, where a hole that could swallow a small car is waiting. 'Back paddle!' Eric yells. And we do, like a well-oiled (but frantic) machine. Then we power ahead into a series of crashing waves. And then it's over – our last rapid.[6]

There are, of course, many Latin American destinations and tourism systems for which the 'natural environment' is of marginal significance to the tourist. The continuing appeal of the Mexican border towns is scarcely rooted in their physical beauty; instead, in Tijuana 'more than anything else there are shops offering a profusion of cheap trinkets (and piñatas of Winnie the Pooh) to tourists who do not seem to either expect or want anything else'.[7]

Tourism's impact on the environment

Many writers have identified the extensive range of impacts imposed on the environment by tourism (see for example, Chris Ryan's *Recreational Tourism: Demands and Impacts*[8] and Stephen Page and Joanne Connell's *Tourism: a Modern Synthesis*[9]). Any attempt to categorise the processes and conditions by which tourism can put pressure on the environment is risky, and important omissions as well as overlaps between categories are inevitable. The purpose of the following overview is merely to demonstrate the great range of situations and conditions through which tourism influences the environment in the LAC countries.

- *Disturbance of habitats and wildlife behaviour*. This manifests itself at various scales, levels of severity and irreversibility as well as the scarcity and perceived significance of the organisms concerned. At the more devastating end of the spectrum, major resort developments may sweep away or at least substantially modify whole ecosystems. On a more local scale, many tourist activities can put pressure on fragile ecosystems by thoughtless, poorly informed or irresponsible tourist behaviour. Examples include disruption of turtle nesting on beaches, off-road driving, use of powerboats and jetskis in natural areas, and the impact of divers and boat anchors on coral reefs.
- *Wildlife in captivity*. Zoos and aviaries are often used to ensure that tourists' expectations are met in situations where the exotic fauna used to sell destinations are unlikely be encountered in the wild. There is also controversy over the fashionable passion for swimming with dolphins. The 'Encounter Program' offered by Dolphin Discovery at Puerto Aventuras on Mexico's

Riviera Maya promises: 'You will also have the opportunity to hug it, kiss it, let it kiss you on the cheek, and enjoy watching your new friend while it performs a series of amazing behaviors.'[10]

- *Destruction of agricultural land, habitat loss and landscape change*. Physical developments in support of tourism such as hotels, apartment complexes, roads and golf courses are often associated with loss of agricultural land, the destruction of wetlands and natural coastlines, or deforestation and the removal of other vegetation. Such landscape interventions have drastic implications for ecosystems, biodiversity, and soil and hydrological processes. Clearly, the scale and severity of impacts varies. In Mexico, the impact on landscapes and habitats of the mega-resort at Cancún contrasts with the network of community-run rural lodging that has developed in the state of Oaxaca.[11]

- *Visitor pressure, erosion and 'wear and tear'*. While direct visitor pressures can be exaggerated, there are many locations where they are a problem. There has been growing international concern that the growth of mass tourism is endangering the iconic Peruvian Inca site at Machu Picchu. Although UNESCO did not take the drastic step of listing the site as 'In-Danger', reports for the 30th Session of its World Heritage Committee in 2006 expressed concern about pressures including visitor and transport impacts, commercialisation and the risks from landslides and fire.[12] There are cases where a very direct, but destructive encounter with the physical environment becomes an attraction in its own right, as suggested by publicity for Parque das Dunas de Genipabu, Rio Grande do Norte, Brazil: 'imagine sliding down 50 meter-high dunes on board buggies conducted by trained drivers who are able to follow through imaginary trails in the white sand at breathtaking speeds'.[13]

- *Water demand*. Tourist consumption of water is likely to be much higher than that by local residents. Tourists expect showers, 'full service' toilets and pools, and their resort surroundings and golf courses to be green. High levels of water demand may accelerate the depletion of groundwater resources or surface water supplies, in some cases to the detriment of local residents, as in Guanacaste province in Costa Rica, reported in Chapter 3.

- *Sanitation and pollution*. Without adequate investment and regulation, tourism facilities may discharge inadequately treated sewage and waste water, with consequent health hazards by release of pathogens, and the eutrophication of water bodies on the surface or underground.

- *Solid waste*. Solid wastes from tourism facilities can impinge on local environments via poorly managed or illegal landfill sites and tips or discarded plastic and other litter on trails, beaches and in coastal waters. In some respects, the problems could be less evident in well-established mass tourism resorts than in more remote destinations, where conventional waste management systems will be poorly developed or absent.

- *Energy consumption*. Many elements of tourism systems create high demand for power, for air conditioning, water heating, cooking, lighting, refrigeration and circulation of water (as well as transport, as below). In the short term, energy demand from tourism may compete for scarce supplies with local consumers, whereas the pollution impacts of electricity generation have both immediate and longer-term implications.

- *Surface transport*. Transport and tourism are inextricably interrelated. The environmental impacts of the additional surface transport activity generated by tourism include fossil fuel energy consumption, atmospheric pollution (with effects felt at various scales in time and space), land take for transport links and terminals, and growing traffic congestion on inadequate urban and rural road networks. In the LAC countries, the majority of tourist-related road movements are short-range, and closely interwoven with those of local populations.

- *Air transport*. With the exception of shorter-range domestic tourism, the Mexico–US border zone and much of the Caribbean cruise market, tourism in the LAC countries is deeply dependent on air transport. Realistically, the long distances between most of the key destination zones and the major generating markets can only be bridged by air. Yet it is now widely recognised that air transport generates a variety of significant environmental costs, including noise, fossil fuel consumption and emissions of pollutants and greenhouse gases. According to the report 'Aviation and the Global Atmosphere' produced by the Intergovernmental Panel on Climate Change (IPCC) in 1999, 'aircraft emit gases and particles directly into the upper troposphere and lower stratosphere where they have an impact on atmospheric composition. These gases and particles alter the concentration of atmospheric greenhouse gases, including carbon dioxide, ozone and methane; trigger formation of condensation trails; and may increase cirrus cloudiness – all of which contribute to climate change.'[14] (This issue is addressed again later in this chapter.)

Direct, indirect and induced environmental impacts of tourism

The list above is indicative, rather than a comprehensive profile of tourism's environmental consequences. The complexity of this issue can be further suggested by borrowing an approach used to represent tourism's contribution to local and regional economies, in which impacts are identified as 'direct', 'indirect' or 'induced'. Thus 'direct' environmental impacts are those generated by the tourism activity itself – the waste produced by a hotel, the noise of a jet departing with a homebound load of inclusive tour visitors or the erosion of a coral reef by the hands and feet of divers. 'Indirect' environmental impacts might include the energy used to launder the sheets and towels of hotel guests, the loss of wetland habitats converted to produce shrimp for hotel kitchens and poolside barbecues, or the fuel used by fleets of trucks removing the waste from condominium blocks in resorts. The 'induced' impacts refer to the environmental impacts generated at second or third hand by individuals and communities that are at least partly dependent on tourism – almost literally 'what goes on behind the hotel'. The nature and scale of such induced effects will reflect the often highly unequal material benefits that diffuse from tourism into the wider community. The wife of the manager of the bank branch used by local hoteliers drives her air-conditioned SUV along a dual carriageway newly constructed through biodiverse scrubland to purchase shrink-wrapped family packs of branded cola in the edge-of-resort supermarket. Elsewhere, the families of chambermaids and those who trim golf courses may simply be tipping the oil from their battered pick-up into a nearby stream, or spraying pesticide on their half hectares of vegetables, to ensure they get at least some surplus to sell in the market.

Some environmental pressures relating to tourism are generated at all three of these levels of impact. For example, traffic levels in resort areas will increase directly as a result of tourists' travel, indirectly through the movement of tourists' supplies and employees in the sector, while further movements are induced by the broader growth in population, material prosperity and consumption in destination zones.

Such observations also confirm the problems of isolating tourism's impact on the environment from other forces and factors. The further from the tourism 'frontline', the more difficult it is to identify the responsibility that falls specifically to tourism. Thus the shrimp farm may also be serving local middle-class or export markets as well as the visitors' poolside grill, and the cars in the WalMart parking lot belong to an increasingly broad consumer class, only some of whom will derive their disposable incomes from tourism.

Sustainability and tourism

Since its emergence in the World Conservation Strategy produced by IUCN, UNEP and WWF in 1980, the notion of sustainability has gained the same status as an exalted global essential as that enjoyed by 'freedom' and 'democracy'. The fillip given to the ideals of sustainable development set out in 'Our Common Future' by the World Commission on Environment and Development, headed by Gro Harlem Brundtland, assured both international acceptance of the principle and a vigorous debate about its meanings and applications that still continues. The legendary definition of sustainable development as development that 'meets the needs of the present without compromising the ability of future generations to meet their own needs'[15] hints at the potential contradictions of the 'now *and* the future' or 'growth with equity' ideals projected by the mainstream sustainability paradigm. The quest for sustainability faces further challenges in the emphasis on an integrated approach to a formidable array of 'interlocking crises'. This approach requires attention to economic, socio-cultural and environmental problems and solutions. So sustainability is inevitably highly contested, elusive territory – the setting for seemingly unending debate on what, for whom, when, how, how much and more.

The application of the 's' word to tourism was somewhat delayed; it is notable that the index of 'Our Common Future' does not include the word 'tourism'. However, sustainable tourism has made up ground rapidly, with its own very ample literature and debates; inserting the phrase into Google produces 10,400,000 hits.[16] While this electronic ubiquity is merely symptomatic, it indicates the presence of a huge academic literature on the topic[17] as well as a vigorous take-up of the concept by the international tourism industry – at least in terms of aspiration. As with

sustainable development, it is widely accepted that sustainable tourism is concern
with a range of interrelated issues, and with seeking a balance between potentially
conflicting objectives and purposes. The World Tourism Organisation (WTO/
OMT) sets out its 'conceptual definition' of sustainable tourism as in Box 4.1.[18]

The WTO/OMT statement is representative of many other interpretations of
sustainable tourism, not least in appearing plausibly broad-based and inclusive.
Sustainable tourism is portrayed as being mindful of the environment and ecological

---

**Box 4.1 World Tourism Organisation's 'conceptual
definition' of sustainable tourism**

Sustainable tourism development guidelines and management practices are
applicable to all forms of tourism in all types of destinations, including mass
tourism and the various niche tourism segments. Sustainability principles
refer to the environmental, economic and socio-cultural aspects of tourism
development, and a suitable balance must be established between these three
dimensions to guarantee its long-term sustainability. Thus, sustainable
tourism should:

(1) Make optimal use of environmental resources that constitute a key
     element in tourism development, maintaining essential ecological
     processes and helping to conserve natural heritage and biodiversity.
(2) Respect the socio-cultural authenticity of host communities, conserve
     their built and living cultural heritage and traditional values, and
     contribute to inter-cultural understanding and tolerance.
(3) Ensure viable, long-term economic operations, providing socio-
     economic benefits to all stakeholders that are fairly distributed, including
     stable employment and income-earning opportunities and social services
     to host communities, and contributing to poverty alleviation.

Sustainable tourism development requires the informed participation of all
relevant stakeholders, as well as strong political leadership to ensure wide
participation and consensus building. Achieving sustainable tourism is a
continuous process and it requires constant monitoring of impacts,
introducing the necessary preventive and/or corrective measures whenever
necessary. Sustainable tourism should also maintain a high level of tourist
satisfaction and ensure a meaningful experience to the tourists, raising their
awareness about sustainability issues and promoting sustainable tourism
practices amongst them.

Source: World Tourism Organisation (2004) 'Sustainable Development of Tourism:
Conceptual Definition', http://www.world-tourism.org/sustainable/concepts.htm.

processes, the integrity and wellbeing of host communities, the economic welfare and persistence of the tourism industry, while not forgetting to let the tourists enjoy themselves. It is perhaps worth noting how the flavour of the WTO/OMT definition of sustainable tourism development appears to have become more diverse and nuanced (which of course, does not guarantee a parallel shift in the delivery of sustainability). In 2001, the definition had echoed the ubiquitous 'Brundtland' mantra on sustainable development:

> Sustainable tourism development meets the needs of present tourists and host regions while protecting and enhancing opportunities for the future. It is envisaged as leading to management of all resources in such a way that economic, social and aesthetic needs can be fulfilled while maintaining cultural integrity, essential ecological processes, biological diversity and life support systems.[19]

The version visible in 2007 also includes basic acknowledgement of the linkage of tourism to poverty alleviation,[20] the suggestion that tourists should be aware of and responsive to sustainability issues (hence 'responsible'?) and also that sustainability should be sought in all forms of tourism.

The claim that sustainable tourism should be regarded as all-inclusive at least provides a point of differentiation between it and the various allied terms that have emerged with the growth in concern about the environmental and socio-economic impacts of tourism. Besides the increasingly prominent 'responsible tourism' that is the basis of this book, perhaps the highest profile has been achieved by ecotourism.

Ecotourism

It is an inescapable truism that ecotourism has become 'big business', both as a label applied to a vast range of tourism destinations, products and experiences and as the focus of vigorous debate. That ecotourism is a contested issue is evident from the numerous definitions on offer; further use of Google using 'definition ecotourism' produces 486,000 hits (with 4,500,000 for 'ecotourism' alone[21]). Many interpretations of ecotourism focus exclusively on tourism that offers encounters with nature, for example: 'tourism involving travel to areas of natural or ecological interest, typically under the guidance of a naturalist, for the purpose of observing wildlife and learning about the environment'.[22]

The International Ecotourism Society defines ecotourism as 'responsible travel to natural areas that conserves the environment and improves the well-being of local people'. It suggests that both providers and participants should follow six key principles, based on Martha Honey's definition:[23]

- minimise impact
- build environmental and cultural awareness and respect
- provide positive experiences for both visitors and hosts

- provide direct financial benefits for conservation
- provide financial benefits and empowerment for local people
- raise sensitivity to host countries' political, environmental and social climate.

The prolific and widely consulted Planeta website, which offers a wealth of online materials on ecotourism in the LAC countries, reduces the list of essential criteria for ecotourism still further:[24]

- it provides for environmental conservation
- it includes meaningful community participation
- it is profitable and can sustain itself.

The scope and scale of ecotourism in the LAC countries have expanded far beyond the point where it can be accurately depicted or measured. However, an extremely crude indication of the spread of ecotourism throughout the LAC countries can be drawn from further recourse to Google, using the search term 'ecotourism' alongside the country name – see Table 4.1.[25] The market leaders on this basis are Mexico, Brazil, Costa Rica and Peru, while at the lower end are Haiti, Puerto Rico, Suriname and Dominica.

The vast majority of ecotourism activity has appeared within the past twenty years, although Arthur Oyola-Yemaiel demonstrates how the early twentieth-century development of Nahuel Huapi National Park and the resort of Bariloche in Argentina was a pioneer application of ecotourism principles, albeit with an additional geo-strategic purpose:

Here, the concept of ecotourism was established by the need to develop communities in frontier areas while maintaining the natural capital. The city of Bariloche, Argentina is a concrete case of sustainable development achieved

Table 4.1 Prevalence of ecotourism in Latin America and the Caribbean – a rough guide?

Number of hits generated by Google, searching on 'countryname + ecotourism' for each country, as at 24 January 2007

Mexico	1,050,000	Nicaragua	354,000
Brazil	785,000	Honduras	344,000
Costa Rica	743,000	Jamaica	338,000
Peru	573,000	Dominican Republic	267,000
Ecuador	555,000	El Salvador	250,000
Argentina	502,000	Guyana	246,000
Cuba	450,000	Colombia	244,000
Belize	437,000	Paraguay	224,000
Guatemala	431,000	Haiti	204,000
Panama	430,000	Puerto Rico	185,000
Chile	397,000	Suriname	159,000
Bolivia	396,000	Dominica	117,000
Venezuela	381,000		

through committing socio-economic forces to generate a tourism industry in an exceptional and pristine area. Ecotourism contributed to the confirmation of the sovereign rights of Argentina over the border regions with Chile by stimulating the local migration to the national park. It also helped preserve the natural environment at Nahuel Huapi National Park by making nature itself the asset of the touristic enterprise. Ecotourism provided the citizens with a place and means of enjoying leisure time while immersing themselves in nature. It generated a permanent settlement around the national parks and contributed to the well-being of the local populations. The concept and the operationalization of ecotourism was born in Argentina with the formation of the National Parks Service during the first year of the twentieth century. This counters the prevailing theory that ecotourism is a concept of the environmental movement of the postmodern era.[26]

The rapid proliferation of ecotourism has inspired a vigorous and extensive critique in activist and academic circles and also within the tourism industry itself. The word's magic allure has been as compulsive for tourism critics as it has for the tourism consumers of the developed North. It is impossible to synthesise the extensive literature[27] here, although a frequent theme has been the misleading application of the term to tourism products that fail to deliver the implied levels of sustainability and responsibility.

Greenwash and ecotourism lite

The liberal and often unjustified claims for ecotourism have brought wide-spread accusations that many tourism operators and destinations are guilty of 'greenwashing'. While conventional dictionaries have not caught up with this representational misdemeanour, the online source Wikipedia defines it as: 'a pejorative term that environmentalists and other critics use to describe the activity of giving a positive public image to putatively environmentally unsound practices'.[28]

Martha Honey offers a nicely sceptical viewpoint based on her close acquaintance with the evolution of Costa Rica's highly publicised ecotourism industry (which is in fact widely regarded as containing some of the more convincing manifestations of ecotourism):

> Today, everything in Costa Rica seems to carry 'eco' in its name. There is, for instance, 'Eco-Playa' (a typical beach indistinguishable from other grey/black sand beaches), 'Ecological Rent-a-Car' (which rents the same vehicles as Hertz, Budget or Avis), 'eco-gas' (super unleaded), 'eco-musica' (songs with environmental themes), and innumerable ecolodges, ecosafaris and ecological cruises. Many of these tourism enterprises can be categorised as ecotourism 'lite', meaning that the company's green rhetoric far outstrips the reality of its adherence to sound ecotourism principles. The classic example of this in Costa Rica as elsewhere is the growing number of major hotel chains that offer guests the 'eco-option' of not having their sheets and towels laundered every day.

Such sensible but relatively minor environmental innovations are advertised with claims such as 'Keep your towels and help save the world!' The reality is that it is the hotels that are saving sizeable sums on their laundry bills.[29]

The various attempts to establish authentic ecotourism credentials via certification and ecolabelling have also raised concerns about the liberal application of poor-quality green colouring in the name of sustainability. Some such initiatives have sought to distance themselves from such accusations; a prominent example is Costa Rica's Certification for Sustainable Tourism (CST)[30] featured in Box 4.2.

Ecolabelling and certification

Costa Rica's CST scheme is one of the more convincing examples of attempts by LAC tourism organisations to establish visible sustainability credentials through systems of ecolabelling and certification. In essence, an ecolabel is a seal of approval awarded to a service to enable both producers and consumers to identify providers whose practices do not harm the environment and society in which they occur. In tourism, ecolabelling and eco-certification refer to the level of responsibility with which the tour company or hotel carry out their operations, so that all aspects of the tour are arranged on an ethical basis, taking due consideration of the local people, the local environment and the labour rights of employees. Eco-labelling schemes generally offer use of a logo to those service providers which meet the criteria for membership of the scheme.

Given the intention of ecolabelling and certification, it would seem self-evident that such schemes should ideally be widely recognised and limited in number. Worldwide, however, the 1990s saw the proliferation of over a hundred such schemes in the tourism industry; by 2001 sixteen different certification labels were available in the tourism industry of Central America alone. Clearly, such a myriad of schemes presents the tourist with a bewildering array of labels and logos. With such a range, it becomes impossible to distinguish the relative merits of each scheme without a degree of research that is not a normally accepted element of the tourism or even ecotourism experience. Not surprisingly, then, as pointed out in an independent report commissioned by WWF-UK,[31] the confusion caused by this proliferation 'has, in turn, led to a lack of consumer demand for certified holidays'.[32] As the same report adds, 'Furthermore, less than one per cent of businesses have joined up to these schemes. The failure to establish clear brand recognition could seriously undermine the potential of certification to bring about sustainable tourism.' Rosselson has also bemoaned the confusion clouding certification schemes: 'there is currently no comprehensive symbol that is universally recognised by tourists as a guarantee that a tour company is both socially and environmentally responsible'.[33]

In 1997, the United Nations Environment Programme (UNEP) conducted a major international survey in order to identify as many existing or planned tourism ecolabel schemes as possible at a global level. From the replies to 400 questionnaires distributed, 28 were selected for detailed analysis.[34] These included such names as the Audubon Cooperative Sanctuary System, Distintivo Ecoturístico, Ecotur,

Box 4.2   Costa Rica's Certification for Sustainable Tourism (CST)

Concern among those Costa Rican tourism operators who claim genuine progress towards sustainability that other less scrupulous businesses were making opportunistic and spurious claims about their green credentials was a key factor in the emergence of the country's Certification for Sustainable Tourism (CST), produced by the Costa Rican Tourism Institute:

> It is a source of concern that sustainable tourism is a concept which is only beginning to be developed, and that it does not yet have a solid reference base or measurable parameters that are appropriately established and universally accepted. This has permitted a significant number of companies to take advantage of the growth of sustainable tourism in an irresponsible manner and thus incur in what we call 'greenwashing', or a commercialization of a tourism experience that does not comply with the reality of the experience provided to the tourist when he/she reaches the destination. The immediate effect of this unethical behavior is to generate a great deal of mistrust on the part of the consumers about the products being offered, with very serious consequences for the whole industry.

The CST programme claims that by 'providing reliable information on the firms that are really making progress in producing a tourist product that is sustainable' it can overcome the unethical behaviour and greenwashing associated with many ecotourism and certification schemes, as lampooned by Sue Wheat and Tourism Concern in Figure 4.1. The CST differentiates between tourism businesses based on the extent to which they comply with a model of sustainability in terms of four 'fundamental aspects':

- *Physical-biological parameters* – evaluates the interaction between the company and its surrounding natural habitat.
- *Infrastructure and services* – evaluates the management policies and the operational systems within the company and its infrastructure.
- *External clients* – evaluates the interaction of the company with its clients in terms of how much it allows and invites the client to be an active contributor to the company's policies of sustainability.
- *Socio-economic environment* – evaluates the interaction of the company with the local communities and the population in general.

Performance on each of these criteria is rated on a scale of five levels of sustainable tourism achievement.

Source: Costa Rican Tourism Institute (2005) 'Certification for Sustainable Tourism', www.turismo-sostenible.co.cr.

Green Globe, Green Leaf, and National Ecotourism Accreditation Programme. The UNEP report acknowledges the confusion caused by unnecessary duplication and recognises the need to develop measures of the effectiveness of tourism ecolabels. It sees ecolabelling as 'part of the worldwide movement by industry towards self-regulation – a movement which is demonstrating how responsible industry can be towards environmental issues and how voluntary action can stimulate or even replace formal legislation'.[35] In this sense, it repeats one of the mantras of capitalist globalisation, that of self-regulation, regardless of the results of its own survey. In other words, UNEP acts as just one more agency which uncritically promotes voluntary self-regulation seemingly unaware of the discourse and debate surrounding the use of such terms and the effects of the application of such policies.

Although ecolabelling is potentially useful as a tool for assisting both producers and consumers, its current practice suggests that it may be used more as an image-greenwash or as an ideological support rather than a genuine guide for tourists. To overcome such suspicions, the myriad of schemes needs to be reduced, rationalised and co-ordinated; an effective mechanism for monitoring schemes developed; logos awarded for compliance with specified criteria rather than merely a commitment to improve; and certification should cover social, cultural and economic aspects of sustainability as well as environmental performance. Without these improvements, eco-labelling and certification are likely to warrant the type of certificate shown in Figure 4.1.

Work on the rationalisation and co-ordination of schemes began in 2001 when the Rainforest Alliance hosted a feasibility project for a global accreditation body called the Sustainable Tourism Stewardship Council (STSC), with the role of setting international standards for the certification of sustainable tourism organisations. Reporting in January 2003, the Rainforest Alliance recommended the 'establishment of regional networks to encourage dialogue among stakeholders and act as a clearinghouse for certification information'.[36] In response, the Sustainable Tourism Certification Network of the Americas was established in September 2003, with the support of the Inter-American Development Bank. The mission of the Network is

to promote sustainable tourism in the region through:
- strengthening tourism initiatives based on mutual respect and recognition
- joint efforts
- harmonizing systems
- sharing information and experience.[37]

Many of the recommendations from both the UNEP report and from the STSC are progressive and much needed, but the UNEP in particular is steeped in the ideology of the World Bank and the IMF. And like the World Bank it has become skilful in surrounding the rhetoric of neoliberalism at the core of its reports with the caring and socially acceptable language of the INGOs. Whether the STSC has been subject to the same subversion of motive as have the United Nations agencies remains to be seen.

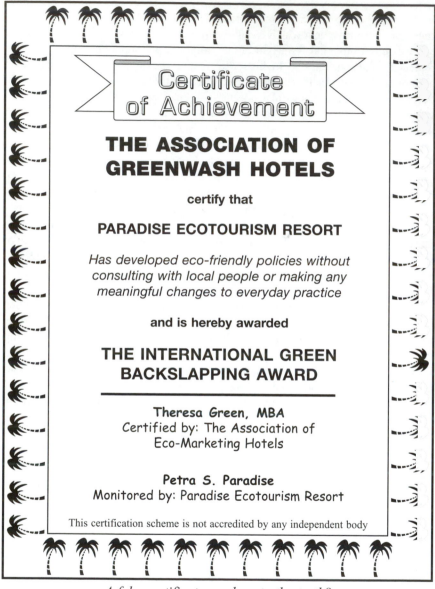

A fake certificate or close to the truth?

Figure 4.1 Certificate of greenwash
Source: Reproduced by kind permission of Tourism Concern.

Ecological footprints, holiday footprints and carbon budgets

Ecological footprinting provides a means of quantifying environmental impacts in a single easily understandable indicator. It also provides a means of identifying opportunities for cost savings and can be adapted for use by tour operators to gauge a 'holiday footprint' for the tours they run. Essentially, this technique measures the impact of a given holiday in terms of the resources it uses relative to a notional annual allowance per person of resources available for use.

It is calculated on the assumption that the earth is a reserve of natural capital, each year producing 'interest' in the form of renewable natural resources such as fish, soil, fresh water and many more. Ecological sustainability requires that we live off this interest rather than eat into the underlying 'capital'. The interest is quantified in units of area. At present there are about two units of area available per person on the planet per year. The WWF-UK estimates that, on a global scale, humanity is annually consuming around a third more resource than the earth produces, which, if accurate, is clearly unsustainable.

Holiday footprinting quantifies the environmental impact of a given holiday according to the measurement of five parameters: the flight, food, energy, waste and 'others'. A report commissioned by WWF-UK and prepared by Best Foot Forward[38] details examples of the application of holiday footprinting as a practical business tool that can be used by tour operators as a guide not only to ways of making holidays more environmentally friendly in terms of the resources used – the 'footprint made' – but also to identify opportunities for cost savings.[39] More recently Colin Hunter and Jon Shaw have applied the technique of ecological footprinting to journeys from five of the G8 countries[40] to a range of Third World countries including four in Latin America.[41] Despite seeking 'conservative estimates of relevance to the ecotourism sector',[42] their findings would not give great succour to the ecotourism propagandists and promoters: 'Despite this, our results suggest that ecotourism holidays involving air travel are likely to produce an absolute demand on global natural renewable resources. The magnitude of this demand may be very substantial.'[43] And later in their work, 'a higher EF [ecological footprint] at the destination area merely reinforces our central conclusion that ecotourism experiences involving international air travel will normally exert an absolute (and substantial) net demand on global natural resources'.[44] Regarding the uncertainty surrounding the oft-debated comparison between the effects of ecotourism and those of mass tourism, Hunter and Shaw comment that 'it would only be logical to infer ecotourism as having a greater impact than mass tourism if, at a global scale, ecotourism products generate more air passenger km. than mass tourism products'.[45]

Hunter and Shaw's work is of considerable significance for the development of ecological footprinting and holiday footprinting, tools of sustainability measurement that may increase in importance in the future. But as they acknowledge in their work the technique needs to be made more robust and at the same time more refined before results from its application can be treated with confidence.

For such a technique to enter into common practice, it would have to become acceptable to the travelling public, for few tour operators and travel agents will begin to describe their holidays in terms of their holiday footprints unless the idea is already widely understood and accepted. Even if such a situation is reached in the future, it would also be necessary to overcome the industry's resistance to change by persuading them that the technique can genuinely work in their financial interests.

Another recently developed methodological tool of possibly greater public acceptability is the carbon budget. The last twenty years have witnessed a growing awareness of the problem of global warming and of the part played in this phenomenon by greenhouse gases, the most significant of which is carbon dioxide, CO_2. As the amount of CO_2 in the atmosphere increases, so the atmosphere heats up. As we have all been taught by the media over the last two decades, we are aware that CO_2 in the atmosphere is increasing largely because of the emissions thereof created by human activity and because of the loss of vegetative material (especially forests) to absorb the CO_2. The science behind this conclusion was confirmed in 2007 by a report of the Intergovernmental Panel on Climate Change.[46]

From the point of view of the tourism industry, global warming is likely to have both positive and negative effects. The negative effects are alarming, especially for specific destinations. Small island states, for example, are particularly at risk from inundation and loss of coral fringes as a result of sea level rises and coral bleaching. Although the threats to small island states in the Pacific or Indian Ocean tend to have the highest profile, the coral cays of the Caribbean Sea are not risk-free in this regard. As some holiday zones such as parts of the Mediterranean become uncomfortable as a result of increasing average summer temperatures and an increasing incidence of malaria and other diseases, other holiday destinations (including some in the LAC countries) may become substitute destinations. The effects of such substitution may be offset, however, by a more positive environmental impact of global warming: namely, the generally warmer summer weather conditions in higher latitudes as well as increases in the real cost of air travel may persuade Europeans to take holidays nearer to home rather than travelling to exotic locations in the LAC countries. Winter tourism is also likely to be affected negatively by global warming, and, whilst Latin American destinations are not famous for this type of holiday, there are ski resorts in the Chilean and Argentinian Andes (one such resort is featured as a case study on the Mapuche Indians in Chapter 5).

Viner and Agnew (1999)[47] report on research from the Hadley Centre[48] which shows the severe impact that climate change could have upon ecosystems and biodiversity in South America. Amazonia is particularly threatened purely as a result of a combination of increasing temperature and decreasing rainfall. They point out that, although tourism in Brazil accounts for only 2.5 per cent of Gross Domestic Product (GDP), compared with a global average of around 10 per cent, the industry is widely considered to be one of the fastest-growing and most promising in terms of the general development of the country. But both nature tourism (still something of a novelty in Brazil) and beach-based tourism are likely to suffer as a result of, respectively, the predicted changes in vegetation and wildlife and the predicted rise in sea levels and consequent increase in shoreline recession around some of the most valued shorefront properties in the country.

Carbon, then, has become an item of relatively common discourse. The idea of measuring an individual's carbon budget may be very recent, but thanks to this discourse, popular understanding of the notion should not present a great obstacle to its use. The 'Flying off to a Warmer Climate?' section of Ben Matthews's Choose Climate website[49] is designed to calculate the amount of greenhouse gas emissions per passenger for any given air journey. The user inputs the origin and destination of their journey, the type of ticket and the type of plane, and the programme calculates the total fuel used per passenger and the total greenhouse gas emissions per passenger. It marks the selected route on a world map and gives the user an idea of the significance of the estimated figures with indices such as its equivalence in the food eaten by an estimated number of people in one year or the size of a typical tree which contains the same amount of carbon emitted by your journey. Other journey parameters, such as the extra cost of the ticket if tax and duty were charged at the same rate as on petrol in the UK, are also estimated; and the global context of the journey is also explained. Output from its website is given in Figure 4.2 for

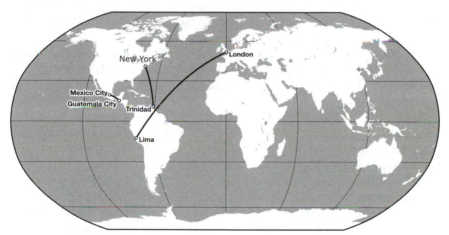

	London to Lima	New York to Trinidad	Mexico City to Guatemala City
Distance (km)	10160	3560	1060
Fuel used per passenger (kg)	730	280	108
Total greenhouse gas emissions per passenger: warming effect equivalent (kg CO_2)	6805	2611	1010
Energy equivalent to electricity used by X 60W light bulbs lit continuously for 1 year, where X=	19	7	3
Contains as much carbon as all the air above X m^2 of the earth's surface, where X=	396	152	59
Contains as much carbon as a typical tree about X m tall, where X=	18	13	10

Figure 4.2 Carbon budget calculations for selected short, medium and long haul flights

Source: Reproduced by kind permission of Ben Matthews who devised and designed the Choose Climate website: www.chooseclimate.org.

an example of short, medium and long haul flights. The following brief extract gives a feel for the spirit in which the calculations work:

> your proposed flight would emit 0.73 tons of carbon (as CO_2) per passenger – i.e., your total sustainable carbon emissions budget for all purposes (including heating, cooking, lighting, local transport, etc.) for 1.82 years. And remember: that's just the CO_2; the total warming effect of $CO_2 + H_2O + NO_x$ is about three times greater.

Ultimately, the questions raised by such websites and commentators is whether, having expanded our horizons through travel and tourism, we should now rein them in so that we may reduce the adverse effects of carbon dioxide emissions on the climate. And if science and society tell us that we must indeed do so, then will we all be allocated a carbon quota (a kind of personal carbon budget) in the future, which we will not be allowed to exceed (unless we have a licence to do so by purchasing another person's unfulfilled quota)? Such an idea may seem a little far-fetched, but in July 2006 the UK Government Environment Minister (David Miliband) revealed his ministry's intentions to examine schemes for implementing and managing tradable personal carbon allowances.[50]

On a scale slightly greater than the personal, Climate Care is a company formed in 1998 'to tackle climate change by reducing greenhouse gases in the atmosphere. We do this by offsetting – making CO_2 reductions on behalf of individuals and companies.'[51] They provide funding for projects that reduce CO_2 in the atmosphere, for example through forest restoration and energy efficiency schemes. The aim of such funding is that it offsets flight emissions. The funding comes from companies which donate to Climate Care a nominal amount for each client who uses services, such as flights, which add greenhouse gases to the atmosphere. A range of transport and tour companies support Climate Care in this way. Two such companies are South American Experience and Journey Latin America, which make donations to Climate Care for every traveller booking packages with them, and which claim that this puts their tours and 'travel onto a sustainable footing'.[52] A similar service is offered by the UK-based CarbonNeutral Company (formerly Future Forests) which helps organisations 'to measure and reduce the CO_2 from some or all of their operation, and then "offset" unavoidable emissions. When emissions are reduced to net zero, the organisation, product or service can carry the CarbonNeutral quality stamp, our registered trademark.'[53]

There is no compulsion on companies to support Climate Care and so it would seem churlish to dispute the motives of the companies that do so. The crux of the argument, however, lies not with the motives of the companies which support Climate Care but with the claim of Climate Care that their funding genuinely offsets the emissions made in the course of air travel. At the same time as its popularity has grown, the exercise of carbon offsetting has increasingly become of dubious value. Tree planting schemes in particular are notoriously unreliable and subject to fire, land takeover and change of use. The *New Internationalist* magazine of July 2006 exposed several carbon offset reforestation schemes which lasted for only short periods of time and managed to offset only a small fraction of the CO_2 emissions caused by the journeys of those who contributed towards the schemes.[54] Criticism of such schemes appears to be mounting (especially among environmental organisations) as their ineffectiveness at combating climate change becomes clearer. The mainstream media, however, is slow to adapt to the critique and still regularly offers advice and possibilities to travellers to offset their flights in articles on 'How to be a responsible tourist'[55] or similar.

Of course it is possible to estimate the tonnage of greenhouse gases absorbed by newly planted forests during their lifetime or to calculate the amount of energy saved through the use of renewable sources of energy – assuming that this same amount would therefore not be produced by other emissions-producing sources. But do these absorptions or savings really offset the emissions produced by the clients of tour companies? Could an absolute reduction in emissions be achieved only by reducing the number of flights sold? Is this approach being used solely as a gesture to the problem – a kind of greenwash? Is it possible to reconcile a tourism industry which encourages people to fly as frequently and as far as possible with attempts to reduce carbon emissions and global warming?

Emissions trading schemes

It is our role here to ask questions rather than to answer them. It is our role also to report the latest developments in techniques which might help us to answer these questions. It is a mark of the recent rise to significance and to public perception of the notion of climate change that in 2005 the UK and European governments began to implement the European Union's Emissions Trading Scheme. The UK's Department of Environment Food and Rural Affairs (DEFRA) explains that

> participating companies are allocated allowances, each allowance representing a tonne of the relevant emission, in this case carbon dioxide equivalent. Emissions trading allows companies to emit in excess of their allocation of allowances by purchasing allowances from the market. Similarly, a company that emits less than its allocation of allowances can sell its surplus allowances. In contrast to regulation which imposes emission limit values on particular facilities, emissions trading gives companies the flexibility to meet emission reduction targets according to their own strategy; for example by reducing emissions on site or by buying allowances from other companies who have excess allowances.[56]

British Airways (BA) has joined the UK Emissions Trading Scheme. The company 'does not accept that the right way to limit emissions is to discourage flying – by punitive taxes or constraints on industry growth . . . if applied to air transport, [this] would lead to extremely negative social and economic effects for the European economy.'[57] However, the company no longer points out, as it used to, that international aviation emissions are not included in the agreed Kyoto caps, and therefore that the scheme includes only its domestic services and UK ground energy sources.

Even if emissions trading were to be extended to international travel, it can be argued that such schemes effectively sanction the continued emission of pollutants into the atmosphere. The perpetrators are given the opportunity to avoid taking action to reduce their own emissions; their 'purchase' of the clean record of others as a counterbalance to their own dirty record cleanses their conscience for emitting so many pollutants themselves. But to 'offset' their own high emissions in this way does not imply that they are *reducing* them, which is widely perceived as the only solution to global warming. Whether these arguments against emissions trading are justifiable is open for debate; certainly proponents of trading would argue strongly that this is a major new technique which allows the world to begin to combat global warming.

These arguments are not raised here in order to cast doubt on the motives and meaning of the companies concerned; rather, they are presented in order to illustrate different facets of the debate. Moreover, they raise again some of the points made in other debates involving the travel and tourism industry, especially those concerning corporate social responsibility, the use of codes of conduct, voluntary self-regulation, ecolabelling and certification (see Chapter 2).

Nature conservation and protected areas

A fundamental instrument for conserving important natural habitats, ecosystems, species and landscapes is the legal designation of protected areas. Reflecting the immense diversity and splendour of the LAC countries' natural environments, substantial areas of the continent have been afforded such protection, as national parks, biosphere reserves and natural parks – at least on paper. National parks have a surprisingly long history in certain LAC countries. In Mexico, for example, the first national park (Desierto de los Leones) was designated in 1917, with many others created in the 1930s. By 2007, Mexico had 67 national parks covering a total of 1.46 million ha. The total area of protected land in Mexico was 22 million ha (11.2 per cent of the national territory), mostly consisting of biosphere reserves and 'areas for the protection of flora and fauna'.[58]

The range and scale of protected areas in the LAC countries continued to expand through the twentieth century and into the twenty-first. Between 1950 and 2003, the area of protected land rose from 17.5 million to 397 million hectares, accounting for 19 per cent of the region's land area.[59] There are plans for further 'megaparks' and biological corridors. Some recent initiatives could give the impression of a dramatic ongoing adoption of the cause of nature conservation, as in the case of the Amazon Region Protected Area (ARPA). This enterprise was launched in 2002 by WWF, the World Bank, the Global Environmental Facility (GEF) and the Government of Brazil as 'the largest, most ambitious effort ever made to safeguard tropical forests'. The scheme 'creates a system of well-managed parks and other protected areas encompassing some 193,000 square miles – an area surpassing in size the entire US National Parks System'.[60] Two new protected areas – 'nearly twice the size of Massachusetts' – have recently been designated within Brazil – the Terra do Meio Ecological Station (8.3 million acres) and the Serra do Pardo National Park (1.1 million acres).

The array of protected areas on official maps is impressive and, admittedly, many areas are better shielded from excessive disturbance as a result of their protected area status. But in many cases in the LAC countries the real extent and quality of protected areas falls well short of the ideals behind designation. In many cases, insufficient resources have been allocated for effective management, so the capacity to resist 'inappropriate' use is often rather slender. Some ostensibly 'protected' areas are little more than so-called 'paper parks', where encroachment by colonists, loggers and other land users threatens the integrity of the nature conservation value associated with the areas. This problem occurs at various scales and in contrasting locations. In Colombia, there has long been concern about the integrity of the country's national parks, many of which are large and in remote areas. An example is the Cahuinarí National Nature Park in the Colombian Amazon.[61]

> Cahuinarí has received heavy pressures from trappers and traders of caiman, cats, otters, and primates. As these populations decrease, poachers are not discouraged, but instead change focus to different species. Intensive commercial fishing for consumption and decoration is also a great threat for fish, crocodiles, and possibly most devastating, the giant river turtle. Increased colonization has come with the discovery of gold on the Brazilian border. A more recent concern has become deforestation in the area around the park.[62]

There are also widespread reports of the threats to Colombian protected areas arising from the illegal drugs industry that is so prevalent in the country. It is claimed that large areas of forest are cleared or disturbed in connection with the plantation of coca, while the attempts at eradication by aerial spraying of herbicides, so strongly promoted and financed by the US government, has deadly consequences for human residents and natural ecosystems.[63]

Designation of land for protection in the interest of nature conservation in the LAC countries has often led to resentment and conflict from indigenous communities, who may have used such 'natural' areas for hunting, gathering, fishing and low-intensity farming for generations. Frequently, parks and reserves have been designated in 'top-down' style, with little effective consultation or participation on the part of local communities. They experience exclusion and loss of freedom and flexibility in terms of access to resources, yet get meagre direct benefits from tourism in return. The potential for conflict may be aggravated because designation of protected areas has taken place in a context of socio-economic dynamism and change, in contrast to the static, 'timeless' conception of indigenous communities. In many cases, local populations have had extensive external contact and influence, fostered by basic education, mobility and access to communications media. As a result the value of forest resources and land may rise, not only because of falling mortality and the subsistence demands of a growing population, but also because local communities have become deeply enmeshed in an economy that offers access to consumer goods and services in return for cash. The growing incentives for more intensive exploitation of forests for timber or for the clearance of land to raise cattle or marketable crops collide with policies favouring nature conservation and protection.

While, in many cases of tension between local communities and conservation policies, tourism is still a relatively minor ingredient, the powerful lure of 'nature', as embodied in the very concept of ecotourism, has inevitably brought more direct conflicts. A number of such cases have been reported in Mexico, including those of the Monarch butterfly reserve in Michoacan state reported in Box 4.3[64] and the

---

Box 4.3   The Monarch butterfly in Mexico

Although a highly distinctive case, David Barkin's account of the Monarch butterfly 'sanctuary' in western Mexico symbolises the socio-cultural contradictions and challenges associated with nature-based tourism.

The full story of the astonishing migration of Monarch butterflies from Canada and the USA to overwinter in a relatively specific area of Michoacan state was only brought to the attention of the wider world in the mid-1970s. The spectacle of the Monarchs quickly attracted large numbers of visitors, and was paralleled by moves to 'protect' their woodland habitat from encroach-ment and damage by local populations through the declaration of a biosphere reserve surrounded by buffer zones, within which resource-based activity was constrained. The enforcement of this resource regulation in the Monarch Reserve generated resentment among local populations, who interpreted conservation measures as largely counter to their interests. Visitor numbers had risen to 250,000 by 1998–99, yet the benefits of tourism were perceived as inadequate compensation for the disruption to livelihoods associated with the reserve, and were concentrated in a four-month 'butterfly season'.

Barkin highlights the need for more diverse forms of tourism as well as broader-based rural economic regeneration, with the benefits spread more widely in both time and space. Interestingly, he supports the expansion of tourism in the area – but with a shift in emphasis to catering for the legitimate recreational needs of less prosperous Mexicans rather than the national and international elites that tend to be drawn to the Monarch Reserve during the butterfly season.

Sources:
David Barkin (1996) 'Ecotourism: A Tool for Sustainable Development in an Era of International Integration?' *Yale School of Forestry and Environmental Studies Bulletin Series* 99: 263–73.
David Barkin (2000) 'The Economic Impacts of Ecotourism: Conflicts and Solutions in Highland Mexico', in P. M. Godde, M. F. Price and F. M. Zimmerman, *Tourism and Development in Mountain Regions*, New York: CAB International, chapter 8.

biosphere reserves of the Yucatán peninsula.[65] And the case of the Galápagos Islands – see Box 4.4 – highlights this tension and the recent notion of 'mass ecotourism'.

Barbara Fraser[66] reports how, in Peru, some of the earlier contradictions and conflicts between the interests of nature conservation and the rights of local communities have been eased by 'new ways of viewing protected areas, clearer demarcation of indigenous lands and efforts to involve communities in park planning'. Fraser cites examples of how local people have been more effectively involved in delimitation and designation of new protected areas, for example in the Otishi National Park on the Apurímac River. The park is flanked by communal reserves where indigenous inhabitants can hunt, fish, gather plants, timber and grow crops. Designation offers protection from invasion by the oil industry to both the core area of great biodiversity and the communal resource areas.

But Fraser also reports concerns about further conflict between nature conservation and indigenous communities, for example in the largest Peruvian national park, Manú. Here, the resident population, while currently very low, is growing rapidly. Indigenous communities are demanding title to areas of land inside Manú Park, which could start to fragment the area protected.

How far is mass tourism the main problem?

There is a widely held assumption by those critical of the industry in the LAC countries that 'mass tourism' is especially damaging and intrusive. The conventional view is that mass tourism is the mainstream villain – guilty of smothering once empty coastlines with high rise blocks full of identical air-conditioned hotel rooms and apartments, strung along traffic-choked boulevards. Apart from planted palms and beaches that may or may not be artificial and laundered nightly, Nature has been vanquished. Here, the hedonistic hordes generate immense quantities of solid and liquid waste that may or may not be returned to the stricken environment, and pour out noise and light pollution. The cornucopia of alternatives – ecotourism, responsible tourism, nature tourism, green tourism *et al.* are projected as more sustainable and generally more environmentally and culturally sensitive than the closely packed hotel and apartment towers in major resorts.

Yet how far is this orthodox critique entirely valid? Some possible environmental *advantages* of mass tourism, as represented in mega-resorts such as Cancún, Acapulco or Mar del Plata could include the following.

- Mass tourism, almost unavoidably, takes place in locations that have already been despoiled: the wetlands have been drained, lagoons polluted and biodiverse scrubland covered in concrete – it was done decades ago, the disasters are in the past.
- Tourists in mass tourism resorts are often spatially 'contained' and also potentially more controllable. They may be more content to stay within 50 metres of the bar or beach and less obsessed with driving their hired jeeps or trail bikes down remote and pristine country roads. For some visitors to the

Box 4.4   Carrying capacity and conservation of the Galápagos Islands

- In 1976 Gerardo Budowski saw the original relationship between tourism and environmental conservation in the Galápagos as one of ideal, mutually supportive symbiosis.[1] In 1976, however, the number of visitors to the islands was only just over 6,000.
- An increased air service led to a doubling of numbers to 12,000 in 1978, at which point the Ecuadorean government 'set a "sustainable limit" to such visitors of 12,000 a year, but for economic reasons then raised the figure to 50,000 a year'.[2]
- In March 1998, the Ecuadorean National Congress passed legislation to promote conservation, protect biodiversity and to establish a Marine Reserve to be managed by the Galápagos National Park Service in collaboration with local institutions.[3] This gave conservationists cause to celebrate, but their celebrations were short-lived.
- In 2001, the islands received 60,000 visitors,[4] . . .
- And in 2003 visitor numbers reached 95,000.[5]
- The Sea Shepherd Conservation Society warned in 2004 that 'We are witnessing the Hawaiianization of the Galápagos'.[6]
- In 2005, visitor numbers climbed to 125,000 with the arrival of 'supersize' cruise liners.[7] Leonor Stjepic, director of the Galápagos Conservation Trust, said 'The Galápagos are not suitable for this kind of mass tourism.'[8]
- In 2006, the World Heritage Committee of UNESCO threatened to put the Galápagos Islands on its 'List of World Heritage Sites in Danger' having noted a number of 'conservation concerns', including: the continuing rise in the number of tourists, as well as accelerated loss of ecological isolation, high immigration rates, and unsustainable commercial fishing.[9]

Notes:
1   Gerardo Budowski (1976) 'Tourism and Environmental Conservation: Conflict, Coexistence, or Symbiosis?' *Environmental Conservation* 3, 1: 27–31.
2   Jonathan Croall (1995) *Preserve or Destroy: Tourism and the Environment*, London: Calouste Gulbenkian Foundation.
3   Reported in *Tourism in Focus*, Summer 1998, 'Conservationists Celebrate New Law for Galápagos', by Galápagos Conservation Trust, London: Tourism Concern.
4   According to David Blanton, executive director of the International Galápagos Tour Operators Association, a non-profit group whose aim is to protect the islands from mass tourism.
5   According to the Sea Shepherd Conservation Society (www.seashepherd.org).
6   As reported in TIM-Team Clearinghouse Bulletin 'Galápagos under heavy assault', 28 May 2004.

7   As reported by the Galápagos Conservation Trust in *Tourism in Focus*, Autumn 2006, 'Galápagos Cruises Go Supersize', London: Tourism Concern.
8   As reported by David Adam (2007) 'A Rat, Insects and Litter: Delights of Mass Tourism Reach Galápagos', London: *The Guardian*, 10 January.
9   UNESCO World Heritage Committee (2006) 'Convention Concerning the Protection of the World Cultural and Natural Heritage', World Heritage Committee Thirtieth Session, Vilnius, Lithuania, 1–16 July, http://whc.unesco.org/archive/2006/whc06-30com-7bE.pdf (accessed 9 February 2007).

LAC countries, some destinations may be imagined as a distinctly hostile environment into which they dare not stray independently, as suggested for instance by a contribution to an online forum concerning the Dominican Republic:

> A friend of mine just came back from staying in Playa Dorada. He told me that his rep said they should be very careful leaving the resort. He has traveled a bit and was skeptical because he was warned: don't take any unofficial tours – yet the only 'official' ones seemed to be the tours offered by his rep. They told him to be careful of local restaurants, something about restaurants that aren't 'crystal.' They told him not to rent motorcycles or cars.[67]

- Mass tourists may be relatively well disposed towards, and also well placed to make use of, public transport, especially in the form of coach or minibus excursions. Besides their reluctance to hire cars and use intimidating local roads, they form a concentrated and thereby viable market for such 'tourism mobility products'. A full coach load can be assembled from just a few pickup points. In contrast, tourists staying at lower densities, in smaller, more isolated self-catering lodges or villas in more remote locations are a much more dispersed market, and are more likely to depend on hired personal four- or two-wheel vehicles, rather than forms of public transport.
- The excessive waste products emerging from modern, high-capacity hotels and condominiums may be more readily disposed of via modern sanitation facilities than the diffused output from lower density, more 'authentic' eco-alternatives. Similarly, mainstream tourism organisations have the resources to invest in energy and water-saving technologies; while their incentive to do so could be seen as 'greenwash', there are also likely to be genuine physical advantages for local environments.
- Larger resorts may generate less social conflict than tourism that offers the prospect of 'living among the locals'. The enclave resort can therefore have advantages in shielding host social environments from excessive tourism penetration. However, if there is an adjacent 'local community', its members may have adjusted fairly well to the madness of tourism – its traumas of change were a generation in the past. In the case of purpose-built resorts such as

Cancún, the 'local community' has, arguably, evolved in close association with the resort, with most resident families having moved into the locality as the resort developed to take advantage of job opportunities.

The counterarguments, however, can be readily identified – not least the sheer scale of major resorts and their level of environmental impacts and resource consumption. The many thousands, even millions of clients who all pour in by plane expect freshly laundered sheets and lavishly watered lawns every day, consume mountains of imported high-calorie foodstuffs in hotels and restaurants and would howl if the air conditioning goes off. Also, 'mass tourism' as a singular phenomenon may be rather elusive on closer examination. Rebecca Torres describes how 'Cancún has been radically transformed form a strictly "sun-and-sand" tourist bubble into a "post-industrial urban tourist space", offering a kaleidoscope of activities'.[68]

While it is therefore unwise to generalise about 'mass tourism' and its possible saving graces, there is perhaps a case for sustaining *existing* tourism centres and activities, and maybe resisting the constant evolution of 'new' tourism places and products. This notion is far from original; the 'honeypot' concept in which tourism activity is concentrated to deflect tourist pressure from more sensitive locations has been applied in many settings and at various scales. A major challenge, however, is that tourism is a dynamic industry that can rapidly spread its sweet allure well beyond the neatly placed honeypot.

The difficulties of containing mass tourism

One of the longest-established examples of tourism concentration has been Mexico's policy of developing large, integrated purpose-built resorts, as exemplified by Cancún, Los Cabos and Huatulco.[69] Concentrated resorts, however, can act as catalysts for the diffusion of tourism development over a wider region. Although Mexico's National Tourism Programme 2001–2006[70] retains the emphasis on mega-resorts such as Cancún, Los Cabos and Huatulco, it also sets out an ambitious vision of spreading and diversifying tourism across wider, more dispersed zones as the basis for regional development. This trend is already well established in the so-called Riviera Maya south of Cancún in Mexico.

The Riviera Maya covers a substantial swathe of Quintana Roo state, along the coast zone south of Cancún. Fast-expanding tourist settlements such as Playa del Carmén and Tulum are part of a strategy for the 'Cancún-Riviera Maya Corridor' that promises such 'benefits' as the addition of 68,000 rooms by 2025 (to produce a total of 113,000), the creation of three new towns and the growth of existing settlements, a new airport, roads and even a railway through the coastal tourism belt.[71] The Riviera Maya concept can in part be interpreted as a drive to sustain the vitality and market share of tourism in Quintana Roo in a competitive global tourism market by offering new products in the form of higher-quality resort environments, accommodation and experiences. Yet in so doing, it has required environmental and social change and intrusion on a scale that seems far from notions of 'sustainability'.

The Riviera Maya forms a much more varied tourist landscape. According to promotional material produced locally,[72] 'Five years ago it was virtually unknown, just a neighbourhood of sleepy fishing villages tucked along the pristine coastline between Playa del Secreto and Punta Allen. Today, the Riviera Maya is the fastest growing tourist destination in the Caribbean region.' The publication stresses the 'wide, sweeping beaches', some of which are 'secluded and solitary', where the only footprints are those of turtles. The area is now 'an explosion of color and diversity'; thus the Riviera Maya is contrasted with Cancún.

> Don't expect high-rise hotels or hustle and bustle like nearby Cancún. Everything here is laid back and relaxing. Even the resorts sprawl lazily along the sun-kissed shore, sometimes hidden within a forest of coconut palms. All of the properties do their best to reside in harmony with nature . . . This philosophy of keeping everything natural is part of what makes the Riviera Maya special.[73]

In this setting, 'adventure-seekers can trek in pristine jungles, photograph exotic wildlife, swim with dolphins, snorkel in natural aquariums, kayak through primitive waterways in the famed biosphere reserve, and explore the finest eco-archaeological theme parks in the world'.[74] Thus there are both claims for the presence of 'pristine', wholly untouched 'natural environments' and promises that the visitor will engage with these environments – snorkelling, photographing, swimming with dolphins or going on 'jungle treks'.

Such claims for environmental harmony and coexistence with untouched nature, however, can be called into question. In the case of so-called 'eco-archaeological theme parks' such as Xcaret and Xel-há, the former projects itself as an ecotourist destination, yet it has many characteristics that critics of ecotourism regard as the antithesis of environmentally and socio-culturally sustainable tourism. The *Lonely Planet Guide* to Mexico describes Xcaret as 'a heavily Disneyfied ecopark. There are still Mayan ruins on the site, but much of the rest has been created or altered using dynamite, jackhammers and other terraforming.'[75] Various claims in the Riviera Maya promotional material noted above also attract a certain degree of scepticism – for example: 'Many resorts flaunt handsome woodwork featuring beautiful hues of varied woods', while the 'Paraíso Tucán' advertises that 'foreigners can own property in Mexico – in this case "an affordable ecological subdivision, legally divided into 2.3 to 19 acre lots, 5 minutes from a spectacular beach"'.[76]

Thus while one of the more blatant forms of 'environmental pressure' – the aesthetic offence of high-rise towers in resorts such as Cancún – is triumphantly discarded as both archaic and ugly, many newer environmental questions are posed. What are the energy and resource demands (footprints) of all these more 'subtle', diverse, dispersed tourism systems as represented by the Riviera Maya? What are the impacts of air-conditioning, transport costs and use of timber? What are the ecological impacts of consuming so much land, despite all the emphasis on landscaping? Although 'some resorts are almost hidden by gardens of rare tropical

plants and flowers, others are tucked away between clusters of coconut palms'.[77]
How far are these heavenly groves in fact highly selective, tamed 'nature'?

How far are 'foreign' tourists to blame? Domestic tourism in LAC and the environment

The dominant perspective on the negative environmental (and socio-cultural) impacts of tourism in the LAC countries emphasises the complicity of tourists from outside the continent. Essentially it is the rich citizens of the North who inflict damage on rainforests, coral reefs and water resources of the South. However, such a stereotype tends to neglect the emergence of substantial tourism systems and markets within the LAC countries. The development process has created a significant (although minority) Latin American middle class who have become enthusiastic consumers of tourism and recreation.

This emergent domestic tourism sector in the LAC countries is diverse and extensive. In many cases it coincides and intersects with incoming foreign tourism. Examples include major Mexican coastal resorts such as Cancún, Acapulco and Puerto Vallarta, high-profile natural sites like the Iguaçu Falls on the Paraná River between Argentina and Brazil and the Monarch butterfly 'reserves' in Michoacan, Mexico. But there are also many tourism destinations and activities in which domestic LAC tourists dominate, such as the beach resorts of Santa Catarina state in Brazil[78] and Mar del Plata in Argentina, and the Andean ski resorts of Argentina and Chile. The rural hinterlands of major cities are increasingly used for recreational tourism activities. In many cases, these are focused on established rural resort zones, such as Ixtapán de la Sal and Valle de Bravo in Mexico, but the more adventurous and mobile urban dwellers are venturing into the wider countryside. Following well-established practice in the North, rural tourism is promoted as a means of economic diversification. Examples include Argentina's Ratur (Red Argentina de Turismo Rural), which offers 'more than 70 destinations in the Argentinian countryside', and the provision of farm accommodation and recreational activities on Colombian coffee farms, including a coffee theme park with robotic dancing orchids – viewed almost entirely by Colombians – see Box 4.5.

While these diverse domestic tourism systems offer economic benefits as well as the potential to raise environmental awareness, they also pose growing challenges. New tourism complexes and their attendant services imply the development of rural land for accommodation, leisure facilities, roads and other infrastructure. As the mobile domestic tourism market expands and diversifies, rural areas are faced with increased road traffic and attendant pollution and congestion, as well as pressures arising from off-road active pursuits such as horse-riding, mountain biking – see Box 4.6 – and off-road driving. As Box 4.6 indicates, there is clear potential for environmental conflict, as the 'new rural tourism' in the hinterlands of major LAC cities is drawn to attractive (and potentially sensitive) landscapes. This tendency is apparent in the vicinity of Mexico City, where forested mountain zones to the south and west of the conurbation areas such as Ajusco and Desierto de los Leones-La Marquesa have suffered encroachment by (often illegal)

Box 4.5   Colombian coffee tours

Jorge Ramírez Vallejo describes how 'in addition to promoting rural tourism through rural stays, theme parks have sprung up in the coffee-growing area to entice families and other tourists. The first and most important of these theme parks is in Montenegro, the so-called Coffee Park or Parque del Café – a combined museum, amusement park, and ecological trail, that provide the tourist with a magical re-encounter with the coffee culture. The park is history, with its life-scale replica of a coffee town; it is nature, with its jungle walk [with] thousands of types of flora and fauna; it's a botanical fantasy with its Orchid Show, in which orchids of every type dance and sing in Disney fashion to entertain and convey ecological messages; it's a food-lover's paradise with countless stands selling cotton candy, ices, and the typical corn pancakes known as *arepas*. The park, which also features a skyride and an observation bridge from which one can overlook the entire state.'

Source: Ramirez Vallejo, J. (2002) 'Agricultural Tourism: Economic Diversification', ReVista Harvard Review of the Americas, Winter 2002, http://www.fas.harvard.edu/~drclas/publications/revista/Tourism/vallejo.html.

---

recreational-residential development, as well as fire, erosion and habitat disturbance as a result of their popularity with Mexican visitors.

How far is tourism to blame? Putting tourism's impacts on the environment in Latin America in perspective

The dauntingly long list of ways in which tourism can threaten LAC environments is widely documented and publicised. But how readily can concerned observers make judgements about the *relative* severity of the problems described? While not wishing to deny or understate the challenges posed by tourism, perhaps it is valid to reflect on comparisons with other development processes, and also to consider the particular places and conditions in which tourism is operating.

The logging and oil industries continue to put pressure on forested environments in the LAC countries; in comparison with these forms of resource exploitation, even the more discredited variants of ecotourism may seem relatively benign. Park planners and communities are now 'uniting against a series of common enemies: oil companies whose concessions overlap both nature reserves and indigenous territories, settlers who migrate from the highlands to clear rainforest for farming, and illegal loggers who are pushing deeper and deeper into the jungle in search of valuable timber'.[79] Local communities, however, may be tempted to collaborate with commercial logging firms, given how lucrative the timber trade can be in comparison with other sources of income. Barbara Fraser suggests that 'after

---

Box 4.6   Mountain biking in Mexico

An example of the off-road tourism activities can be found in the Desierto de los Leones – La Marquesa area on the Toluca highway west of Mexico City, described as:

> the mecca of mountain bikers in Mexico City . . . The main trail that leads to many other destinations starts off as a long moderate climb on a fire road, and all the way has . . . absolutely everything, single track, downhill, climbs, you name it, by far the most popular trail or route to do down here, and just 15 minutes away from the city, excellent weather and riding conditions all year long, highly recommendable.

However, such peri-urban wilderness environments are still partly protected by potential users' real and imagined fears about security and hostile locals:

> Chiluca used to be fun. There were many trails and the MTBkers respected locals and the environment. Since the entry of motorbikers, the place has spiralled downwards. There are too many 2-cycle freaks, Tony Soprano narco-style, who carry handguns inside their body armor. Beware of these people. Also, thanks to the large number of motorbikers, the residents of Chiluca have put up storm fencing and eliminated many single-track trails.

Source: MTB Review.com (2002) Review of Chiluca-Xinte Estado de Mexico trail, content.mtbr.com/trails-around-the-world/trails-latin-america/PRD_168063_4550 crx.aspx, accessed 21 March 2005.

---

expenses, a family doing small-scale, informal logging can earn about \$800 a month'.[80]

In a very different setting, it may be that, in major LAC cities, the total impact of tourism in terms of environmental pressures such as air pollution, water demand and traffic generation may be relatively modest. The activities of local residents and businesses are likely to be overwhelmingly more significant. But the picture is clearly different in smaller, more specialized tourist resorts in rural and coastal areas. Yet although it is valid to put tourism's impacts into perspective, tourism clearly reflects and operates within the same prevailing model of capitalist economic development that propels these coincident sources of environmental damage.

Notes

1 Patagonia-Argentina.com (2004) 'Ushuaia, the Southernmost City in the World', www.patagonia-argentina.com/i/tierradelfuego/ushuaia/ushuaia.htm, accessed 23 February 2005.
2 *Ibid.*
3 Latin Trails website, latintrails.com/destinations/bolivia/uyuni%20salt%20lake.htm, accessed 26 July 2004.
4 Macaw Landing Foundation (2001) 'Ecotourism: Paititi', www.cnnw.net/~mlf/paititi.html, accessed 24 February 2005.
5 *Ibid.*
6 Paul Goldsmith (2001) 'Fighting the Fu', *American Way*, 15 February 2001, www.americanwaymag.com/travel/feature.asp?archive_date=2/15/2001, accessed 22 February 2005.
7 Forrest D. Colburn (2002) *Latin America at the End of Politics*, Princeton: Princeton University Press, p. 68.
8 Chris Ryan (2003) *Recreational Tourism: Demands and Impacts*, Clevedon: Channel View Publications.
9 Stephen Page and Joanne Connell (2006) *Tourism: A Modern Synthesis*, Second edition, Andover: Thomson.
10 Dolphin Discovery, www.dolphindiscovery.com accessed 7 February 2007.
11 Ron Mader (2004) 'Tourist Yu'u Villages, Oaxaca', gorp.away.com/gorp/location/latamer/mexico/mader2.htm, accessed 6 February 2005.
12 UNESCO World Heritage Committee (2006) Convention Concerning the Protection of the World Cultural and Natural Heritage, World Heritage Committee Thirtieth Session, Vilnius, Lithuania, 1–16 July, http://whc.unesco.org/archive/2006/whc06-30com-7bE.pdf, accessed 6 February 2007.
13 Folha de Turismo (2002) 'The Dunes of Genibapu: Emotion and Beauty', *Folha de Turismo* Special edition WTM, London, 11–14 November, p. 13.
14 Intergovernmental Panel on Climate Change (1999) Aviation and the Global Atmosphere, Summary for Policymakers, www.grida.no/climate/ipcc/aviation/004.htm, accessed 20 June 2005. For further coverage of the environmental impacts of air transport see also: Paul Upham *et al.* (2003) *Towards Sustainable Aviation*, London: Earthscan; John Whitelegg and Nick Williams (2003) *The Plane Truth: Aviation and the Environment*, London: Ashden Trust and Transport 2000.
15 World Commission on Environment and Development (1987) *Our Common Future*, Oxford: Oxford University Press, p. 8.
16 Based on entering 'sustainable tourism' into Google, as at 2 February 2007. The 10,400,000 resulting hits are admittedly overshadowed by the 75,900,000 achieved by using the search term 'sustainable development'!
17 This immense literature can only be symbolised here, for instance by the *Journal of Sustainable Tourism*, and texts such as: France, L. (ed.) (1997) *The Earthscan Reader in Sustainable Tourism*, London: Earthscan; Hall, C. M. and Lew, A. (1998) *Sustainable Tourism: A Geographical Perspective*, Addison Wesley; Swarbrooke, J. (1999) *Sustainable Tourism Management*, CAB; Harris, R. *et al.* (2002) *Sustainable Tourism: A Global Perspective*, Oxford: Butterworth Heinemann; Pineda, F. D. and Brebbia, C. A. (eds) (2004) *Sustainable Tourism*, Southampton: WIT.
18 World Tourism Organisation (2004) 'Sustainable Development of Tourism: Conceptual Definition', http://www.world-tourism.org/sustainable/concepts.htm, accessed 6 February 2007.
19 Cited in Liu, Z. (2003) 'Sustainable Tourism Development: A Critique', *Journal of Sustainable Tourism* 11, 6: 460.
20 The WTO/OMT's self-declared commitment to encouraging 'pro-poor' tourism is represented by the Sustainable Tourism-Eliminating Poverty (ST-EP) initiative

launched in 2003. However, the prominent slogan on the ST-EP web page (www.world-tourism.org/step/step.htm) may betray its roots rather too transparently: 'Liberalization with a human face' – see Chapter 2. The core objective of ST-EP is declared to be 'raising living standards through expansion of the (tourism) sector'.

21 Based on entering 'definition ecotourism' and 'ecotourism' into Google, as at 30 January 2007.

22 Answers.com: 'ecotourism', www.answers.com/ecotourism&r=67, accessed 21 March 2005.

23 The International Ecotourism Society (2004) What Is Ecotourism? www.ecotourism.org/index2.php?what-is-ecotourism accessed 26 February 2005. See also Honey, M. (1999) *Ecotourism and Sustainable Development: Who Owns Paradise?*, Washington DC: Island Press, and Merg, M. (2007) *Defining ecotourism*, Washington DC: Untamed Path, www.untamedpath.com/Ecotourism/defining.html, accessed 7 February 2007. It is interesting to note that between 2005 and 2007 a seventh key principle – 'support international human rights and labour agreements' – had been omitted from the The International Ecotourism Society's recommendation.

24 Mader, R. (2007) 'Exploring ecotourism: definitions', Planeta.com, www.planeta.com/ecotravel/tour/definitions.html, accessed 6 February 2007.

25 As per 24 January 2007.

26 Oyola-Yamaiel, A. (1997) 'Conservation and Ecotourism in Argentina: The Development of Bariloche and the Formation of the Nahuel Huapi National Park', Planeta.com, www.planeta.com/planeta/97/0597argentina.html, accessed 21 March 2005.

27 Some prominent representatives of the literature on ecotourism include: Cater, E. (2004) 'Ecotourism: Theory and Practice', in Alan Lew, Colin Hall and Allan Williams (eds), *A Companion to Tourism*, Oxford: Blackwell; Ceballos-Lascurain, H. (1996) *Tourism, Ecotourism and Protected Areas: The State of Nature-based Tourism around the World*, Gland. IUCN; Duffy, R. (2002) *A Trip Too Far: Ecotourism, Politics and Exploitation*, London: Earthscan; Fennell, D. (2003) *Ecotourism: An Introduction*, Second edition, New York: Routledge; Honey, M. (1999) *Ecotourism and Sustainable Development: Who Owns Paradise?*, Washington DC: Island Press; Page, S. and Dowling, R. (2002) *Ecotourism*, Harlow: Prentice Hall; Caroline Stern, James Lassoie, David Lee and David Deshler (2003) 'How "Eco" Is Ecotourism? A Comparative Case Study of Ecotourism in Costa Rica', *Journal of Sustainable Tourism* 11, 4: 322–47.

28 Wikipedia (2007) Greenwash, http://en.wikipedia.org/wiki/Greenwashing, accessed 7 February 2007.

29 Honey, M. (2003) 'Giving a Grade to Costa Rica's Green Tourism', *NACLA Report* XXXVI, 6, May/June: 44.

30 Costa Rican Tourism Institute (2005) Certification for Sustainable Tourism, www.turismo-sostenible.co.cr/EN/home.shtml, accessed 20 June 2005.

31 WWF-UK (2000a) 'Tourism Certification: An Analysis of Green Globe 21 and Other Certification Programmes', Godalming: WWF-UK.

32 WWF-UK (2000b) 'Tourism Certification Schemes Still Leave Much to Be Desired', 29 August, www.wwf.org.uk/news/n_0000000132.asp.

33 Rosselson, R. (2001) 'Ethical Tourism', *Ethical Consumer* 69: 28–9.

34 United Nations Environment Programme (1998) *Ecolabels in the Tourism Industry*, New York: UNEP.

35 *Ibid.*, p. 5.

36 Rainforest Alliance (2005) 'Sustainable Tourism Certification Network of the Americas' www.rainforest-alliance.org/programs/tourism/certification/network-of-americas.html, accessed 12 July 2005.

37 Rainforest Alliance (2007) www.rainforest-alliance.org/tourism.cfm?id=network, accessed 7 February 2007.

38 Best Foot Forward is an independent consultancy specialising in natural resource accounting techniques such as ecological footprinting. See www.bestfootforward.com.
39 WWF-UK (2002) *Holidays Abroad Needn't Cost the Earth*, Godalming: WWF-UK.
40 The G8 countries are: USA, UK, Japan, France, Germany, Italy, Canada and Russia. Of these, the five chosen in Hunter and Shaw's work were USA, Germany, UK, Japan and France.
41 Colin Hunter and Jon Shaw (2006) 'Applying the Ecological Footprint to Ecotourism Scenarios', *Environmental Conservation* 32, 4: 7.
42 *Ibid.*, p .7.
43 *Ibid.*, pp. 7–8.
44 *Ibid.*, p. 8.
45 *Ibid.*, p. 9.
46 Intergovernmental Panel on Climate Change (2007) *Climate Change 2007: The Physical Science Basis*, Geneva: IPCC.
47 Viner, D. and Agnew, M. (1999) *Climate Change and Its Impact on Tourism*. Report prepared for WWF-UK and produced by the Climatic Research Unit of the University of East Anglia.
48 Hadley Centre (1998) *Climate Change and its Impacts: Some Highlights from the Ongoing UK Research Programme*, UK Meteorological Office Publication, 12 pp.
49 Choose Climate website: www.chooseclimate.org, accessed 12 February 2007.
50 www.defra.gov.uk/news/2006/060719b.htm – and as reported in Friends of the Earth press release 21 July 2006.
51 Climate Care website: www.climatecare.org, accessed 12 February 2007.
52 South American Experience (2005) Sustainable Tourism: CO_2 Emissions and Climate Care, www.southamericanexperience.co.uk/information/sustainabletourism.html, accessed 12 February 2007.
53 CarbonNeutral Company (2007) www.carbonneutral.com/pages/whatwedo.asp, accessed 12 February 2007.
54 *New Internationalist* 391 'CONNED: Carbon Offsets Stripped Bare', July 2006, Oxford.
55 As in *The Guardian* (2006) 'How to Be a Responsible Tourist', London, 17 July – associated with Esther Addley's article 'Boom in Green Holidays as Ethical Travel Takes Off'.
56 Department of Environment, Food and Rural Affairs (2005) Emissions Trading Schemes, www.defra.gov.uk/environment/climatechange/trading/index.htm, accessed 11 July 2005.
57 British Airways (2007) Air transport and climate change, www.britishairways.com/travel/crglobalwarm/public/en_gb, accessed 12 February 2007.
58 Areas Naturales Protegidas (2005) Comision Nacional de Areas Naturales Protegidas, www.conanp.gob.mx/anp/anp.php, accessed 7 February 2007.
59 FAO (2004) More Protected Areas and Planted Forests in Latin America and the Caribbean, FAO www.fao.org/newsroom/en/news/2004/ 51189/index.html, accessed 21 March 2005.
60 WWF (2005) 'Amazon Region Protected Areas Program (ARPA): A Giant Step to Protect the Amazon Forest' www.worldwildlife.org/wildplaces/amazon/projects/arpa.cfm, accessed 7 February 2007.
61 The Nature Conservancy (2005) 'Cahuinarí National Natural Park, The Nature Conservancy: Parks in Peril', parksinperil.org/wherewework/southamerica/colombia/protectedarea/cahuinari.html, accessed 11 July 2005.
62 *Ibid.*
63 Philip Cryan (2004) 'The Drugs Wars Next Casualty: Colombia's National Parks', *Counterpunch*, 23 March, www.counterpunch.org/cryan03232004.html, accessed 11 July 2005.

64 David Barkin (2000) 'The Economic Impacts of Ecotourism: Conflicts and Solutions in Highland Mexico', in P. M. Godde, M. F. Price and F. M. Zimmerman, *Tourism and Development in Mountain Regions*, New York: CAB International, chapter 8. See also: David Barkin (1996) 'Ecotourism: A Tool for Sustainable Development in an Era of International Integration?' *Yale School of Forestry and Environmental Studies Bulletin Series* No. 99, 263–73.

65 Richard C. Smardon and Betty B. Faust (2006) 'Introduction: International Policy in the Biosphere Reserves of Mexico's Yucatán Peninsula', *Landscape and Urban Planning* 74: 160–92.

66 Barbara J. Fraser (2003) 'Joining Forces for Peru's Rainforests', *NACLA Report on the Americas*, 36, 6: 13–16.

67 DR1 Forums (2003) 'Dangerous to leave resort?' dr1.com/forums/, accessed 24 February 2005.

68 Rebecca Torres (2002) 'Cancún's Tourism Development from a Fordist Spectrum of Analysis', *Tourist Studies* 2, 1: 87–116.

69 Ludger Brenner (2005) 'State-planned Tourism Destinations: The Case of Huatulco, Mexico', *Tourism Geographies* 7: 138–64.

70 Secretaria de Turismo, Mexico (2001) *Programa Nacional de Turismo 2001–2006*.

71 *Ibid.*, p. 149.

72 Fideicomiso para la Promoción Turística de la Riviera Maya (2003) 'Destination Riviera Maya: the Official Visitor Guide and Tourism Publication'.

73 *Ibid.*

74 *Ibid.*

75 *Lonely Planet Mexico* (2003) p. 977.

76 Fideicomiso para la Promoción Turística de la Riviera Maya (2003) 'Destination Riviera Maya: the Official Visitor Guide and Tourism Publication'.

77 *Ibid.*

78 For example, the coastal resort of Balneário Camboriú in Santa Catarina state, Brazil, where 'the conurbation process has led to a rapid and careless evolution of the urbanization process', and the rapid expansion of domestic tourism has resulted in conflicts such as: 'local economic distortions, environmental degradation and loss of cultural identity'. Marcus Polette and Giancarlo Donato Raucci (2003) 'Methodological Proposal for Carrying Capacity Analysis in Sandy Beaches: A Case Study at the Central Beach of Balneário Camboriú (Santa Catarina, Brazil)', *Journal of Coastal Research* 35: 94–106.

79 Barbara J. Fraser (2003) p. 13.

80 *Ibid.*, p. 15.

5 Indigenous peoples and tourism in Latin America and the Caribbean

For tourists one of the key 'selling' points of Latin America and the Caribbean is the colour and diversity of its indigenous human life. From the amazingly colourful costumes of the Mayas in Central America to the bustling markets of the Otavalan Indians of Ecuador, the apparent mystique and exoticism of the many Amazonian groups and the seeming austerity of the Garífuna descendants of African and indigenous peoples on the island of St Vincent, many indigenous peoples are involved in the activity of tourism either as objects to be viewed or increasingly as active and willing participants in the exchange.

The attraction of indigenous peoples to First World tourists is akin to the allure of the 'other', something distinctly different from western lifestyles, and a difference which is heavily romanticised by travel writers, tour operators and 'alternative' holiday advertisements. These 'other' cultures are portrayed as exotic, sensuous, erotic, simple, sustainable, naturally colourful and in tune with nature. In this respect, tourism is the vehicle for the process of 'othering', a process described by Hall,[1] and one through which western tourists try to represent the notion of an authenticity they feel they have lost in their own lives and lifestyles.

The issues raised by the involvement of indigenous peoples in tourism are considered in this chapter. Essentially these issues are: the cavernous gap between a modern westernised lifestyle and an 'unassimilated' indigenous lifestyle; the role that the tourism industry plays in bridging this gap or exaggerating the gap; the notion of authenticity and whose perception of it rules; and the different interpretations of capturing the spirit of a people through photography. But perhaps the phenomenon of tourist visits to indigenous groups raises the whole idea of development and what it means more than any other associated issue.

This chapter begins with an analysis of the place of indigenous groups and their life and cultures in the model of development which now prevails in the LAC countries. This includes an examination of the varying ways in which governments perceive and treat these groups. We also focus on the nature of their involvement with the tourism industry, specifically through the medium of photography. And this is followed by a number of case studies of tourism initiatives which specifically involve indigenous groups. Finally, we give a summary of the issues raised by the exchange between indigenous groups and the tourism industry.

Diversity, assimilation, zooification and transculturation

Within societies which broadly pursue the model of western capitalist economic development, there exist enormous cultural differences, and from the current round of economic globalisation that is identified with this model has sprung a new consciousness of diversity. The chance to view many of these cultural differences is often the principal motive for tourism. Beyond these societies – yet within them – there live tribal peoples whose cultures are hugely at variance with those of the so-called 'developed' world. For the tourism industry such variance represents a set of opportunities: to revisit our anthropological roots; to confront our own species in entirely natural and 'undeveloped' environments; to come face to face with the 'other' (and all that otherness embodies); and to see a reality so different from our own that it must be authentic.

For the tribal peoples themselves this cultural difference from their visitors can also represent a set of opportunities: to control their own contact with the outside; to enter the money economy in a significant way; and to promote in others a general understanding of and sensitivity towards their life, culture, society and belief systems. But it can also represent a series of dangers: of subversion of lifestyle and culture due to the corrupting effect of money; of corrosion of lifestyle as new ways, practices and fashions are introduced without due care and forethought; of exposure to disease; of conflict with squatters and developers; and in the worst of cases of extinction. These are no small dangers.

The issues behind the assimilation of tribal peoples into the predominant culture and economy of the nation state that surrounds them have been debated since the time of the Spanish conquistadors. It would be reassuring to think that the debate is now rather more enlightened than at the time of Pizarro when conquering, pillaging, raping and exterminating were the prevailing forms of contact. As the cartoon in Figure 2.1 (page 15) illustrates and as some of the attitudes exhibited by those in public and private power and some of the case studies presented in this chapter demonstrate, however, the age of enlightenment has still to reach many corners. And it is by no means certain that the contact brought by tourists (and those who cater for them) should be seen in a different light from the contact of the conquistadors with the indigenous populations of their time.

That the policy of extermination was the chosen solution to the 'Indian problem' in the days of the conquistadors is illustrated by estimates of the population of tribal peoples.

> The best recent investigations credit pre-Columbian Mexico with a population between 30 and 37.5 million . . .; Central America had between 10 and 13 million. The Indians of the Americas totalled no less than 70 million when the foreign conquerors appeared on the horizon; a century and a half later they had been reduced to 3.5 million.[2]

Of the Amazon region, Progressio (formerly the Catholic Institute for International Relations, CIIR) states that:

Five million indigenous people ('Indians') are believed to have been living in what is now Brazil when Europeans arrived. Their number now is put at about 250,000. They were wiped out by the Europeans, some through mere contact with diseases to which they had no immunity, some through forced labour on the plantations. Many were killed so that the invaders could occupy their land. Despite laws in the 19th century to protect them, they have generally been regarded as an obstacle to progress, and have survived as communities only in remote areas or reserves.[3]

The slaughter continued up to and through the twentieth century, throughout which Survival International,[4] an INGO which defends the rights of tribal peoples, estimated that Brazil's tribal groups were becoming extinct at the rate of one per year.[5] Obviously, the pre-Columbian indigenous populations of the Americas suffered as a result of the contact with the new economic and cultural system brought from Europe. But factors which made the twentieth century different from the four previous centuries are the rate at which tribal groups have come into contact with the new conquistadors, the rate and scale of development of the land by the new economic system of twentieth-century western capitalism and the rate of their extermination. The exposure of entirely self-sufficient and self-contained tribal groups to a modern market system has exerted forces and stresses of rapid and at times unbearable change on them. As Marcus Colchester has stated: 'For the first time for millenia, the tribals find that they can no longer easily determine their own destinies. Forces beyond their immediate control determine their relations with outsiders and the economic relations they establish with them.'[6]

Throughout the twentieth century in Brazil many of these new contacts were made with squatter farmers, gold miners and cattle ranchers among others; in Ecuador it was the oil men and north American evangelists; in Panamá it was the military, private militias and maritime traders; in other countries there were many of these same colonisers of new areas together with agents of the timber industry, exploiters of mineral wealth, explorers, and many other opportunists. Figure 5.1 identifies the culprits in the eyes of the Indians who wrote the graffiti on this wall in Puyo, Ecuador.

And then, in the later part of the century, came the tourists.

Some tribal groups, admittedly those who had already been contacted and who had survived the experience, seized the opportunity offered by tourism. The Kuna Indians, for instance, take a considerable profit (in total often more than US$10,000) each time a cruise ship calls at one of their islands in the San Blas Archipelago off the Caribbean coast of Panamá. The example of the Kuna is taken up as a case study later in this chapter (p. 152). Others, such as the indigenous peoples of southern Belize, have formed collaborative groupings to control, promote and profit from so-called ecotourism ventures. (Again, the case of the Toledo Ecotourism Association in Belize is taken up later in this chapter – p. 163). Others, however, have been traumatised and decimated by their contact with tourism, and there exist many cases where the experience may not have been traumatic but has certainly been exploitative. Some of these, such as the Mapuche in Chile and Argentina and many

'No to the invasion of Indian lands by oil companies, logging companies, agro-industrial companies, land owners, professionals and land speculators'

Figure 5.1 Indian graffiti in Puyo, Ecuador
Photo: Martin Mowforth

groups in the Rio Negro basin in Barcelos, Brazil, are covered as case studies later in this chapter.

Where extermination was not the deliberate policy, the twentieth century saw a general assumption by governments of the day that tribal groups should be assimilated into the prevailing metropolitan culture. The idea that civil, human, social and land rights should be accorded to tribal groups as separate entities was generally viewed as ridiculous. Autonomy for such groups was entertained as an acceptable idea only on the assumption that they had no rights whatsoever. In short, they were widely viewed by many of those at the frontier of western capitalist development as sub-human. Assimilation could be countenanced only if they accepted their role on the bottom rung of western capitalist society – it is not for nothing that tribal groups are sometimes referred to as the Fourth World.

Third World governments more often than not serve as agents of the prevailing economic model of development, and in that role they are keen to capitalise on the income potential represented by natural resources within their national boundaries. Exploitation of natural resources such as mineral wealth, timber, plant diversity, hydroelectric energy and even wildlife has proved easy to exploit if destruction of the natural environment and removal of its inhabitants can be disregarded. And some governments of the LAC countries have indeed managed to disregard the natural ecosystems in their 'development' of natural resources whilst at the same time waxing lyrical about the need to protect the environment.

In the Oriente region of Ecuador, for example, both the government and the oil companies have spoken of the importance of the preservation of the rainforest whilst overseeing the most obvious and transparent process of its destruction during the 1970s, 1980s and 1990s in their rush to gain revenue and profits. During that time the supposed wealth from oil has seen levels of poverty in the country rise from 50 per cent of the population to over 70 per cent and the country's debt burden soar from a quarter of a billion US dollars to over \$16 billion.[7] The Oriente region is home to eight distinct tribal groups numbering around 200,000 people, many of whom have suffered and are suffering a range of health problems because of the pollution caused by the oil extraction process.[8]

In Brazil, despite the existence of FUNAI, the Brazilian Indian Foundation, successive governments have ignored the problems and even the existence of the indigenous populations and have turned a blind eye to incursions into their territory by miners, land owners, cattle ranchers and others seeking to exploit the vast mineral and natural wealth of the rainforest. Indeed, FUNAI has often suffered reductions in government resources, has a history of corruption,[9] and its government-appointed directors have often carried with them agendas which worked against the needs of the indigenous groups whose affairs it is supposed to represent. In February 1999, for instance, Marcio Lacerda was appointed president of FUNAI. A report by the Indigenous Missionary Council expressed its concern over Lacerda's appointment on the grounds that his approach 'is not focussed on indigenous peoples' cultures and needs but on the natural resources of their territories'.[10] Later in that year, Joelle Diderich described the new FUNAI president as wanting 'indigenous tribes to start paying for themselves after years of relying on government hand-outs',[11] and quotes Lacerda as saying: 'We are talking about the use of natural resources, managed as much as possible by the communities themselves in their own interests.' Although Lacerda went on to acknowledge that such communities cannot be abandoned to play with market forces, it is clear that capitalist competition between tribal groups and transnational companies is most unlikely to be conducted on a level playing field.

In 2002, the Socio-Environmental Institute of Brazil reported that 'the region of the Xingu is still marked by the mentality that *development* is synonymous with cutting down the forest, monoculture, cattle-raising, etc.'.[12] The pressures exerted by the international financial institutions such as the World Bank and IMF to liberalise the Brazilian economy – see Chapter 2 – are only likely to encourage this rush to exploit the Amazon and thereby to increase the pressure on tribal groups to assimilate themselves into the western economic model of development or to disintegrate.

A variety of attempts to achieve assimilation without extinction arose in the latter half of the twentieth century, notably that of the Villas Boas brothers in Brazil. In their attempts to preserve indigenous cultures and to protect tribal groups from extinction through contact with western culture, the four Villas Boas brothers lobbied the government to set aside land for the Xingu National Park within which all means of protection would be afforded to flora, fauna and human populations. In 1961 the reserve was created as an area of 26,000 square km, with Orlando Villas Boas as its director for its first seventeen years, and the Villas Boas brothers persuaded a number of threatened tribes elsewhere in Amazonia to re-locate within its boundaries. The population of tribal groups inside the park is currently increasing.[13] The park still exists and 'is today a green oasis surrounded by devastated forest';[14] but other anthropologists have criticised Orlando Villas Boas as being paternalistic for seeking to prevent all forms of contact by the Indians with the outside world and for turning the Xingu National Park into a showcase.

The dilemma between the policies of preventing external contact and assimilating tribal groups has now moved on. With the globalising reach of today's economy, the former is less of a serious option, and the question is now more one of to what degree and how are tribal groups assimilated. As Michael Astor explains of Indians in the Xingu National Park, 'Satellite dishes sit outside many of the long houses feeding a handful of Brazilian TV channels to generator-powered televisions'.[15] A similar example, that of the Guarani-Kaiowá also in Amazonia but outside the Xingu National Park, cited by Survival International, outlines this strange juxtaposition between primitiveness and modernity:

> there is an obvious taste for some aspects of modern life in the village. One community member who lobbies for the village's rights with local government gets dropped off from town in a car, a mobile phone protruding from his top pocket and a video player tucked under his arm . . . a man is singing to the setting sun in Guarani, shaking his rattle in time to his prayers. But his song is soon drowned out. Brazil's most popular nightly soap opera is starting, its maddeningly kitsch overture wafting from the mud hut next door.[16]

As Mairawe Kayabi, president of the Xingu Indian Land Association, explains, 'All the stuff on the television puts stuff in the young people's heads. They are attracted to whatever comes from outside. This is a cause for a lot of disagreement among the leadership.'[17] Tourism has not yet entered the Xingu National Park, but clearly other external influences have, and perhaps it is only a question of time before tourists and tourism entrepreneurs begin to promote their own form of interaction with the tribal groups therein.

An example which illustrates that tourism can also serve as a source of friction between developers and indigenous groups is found in northern Honduras, where the beaches of the country's Caribbean coastline are a major attraction for tourists. Since the early 1980s the Honduran government has made consistent efforts to attract both tourists and foreign investors in the industry, to the point where the tourism industry is now the country's second largest foreign exchange earner.

The Caribbean coastline of Honduras, however, is also home to many Garífuna communities which have ancestral title to their land. Not all of these titles are recognised by the Honduran government, which in 1998 reformed the Constitution to permit foreigners to acquire land less than 40 km from the coast, an entitlement previously prohibited. The communally held titles grant the Garífuna communities rights to their area in perpetuity, and land may not be sold or transferred to owners outside the community. The constitutional change gave rise to Garífuna fears that big hotel investors would push the Garífuna out of their homes. In 2004 Eva Thorne reported that in some cases these fears had been well grounded:

> the tourism boom of recent years and the consequent demand for valuable beachfront property has created incentives for land invasions and intimidation, as well as bribery and outright violence against Garífuna communities. While other commercial interests have historically threatened encroachment on Garífuna lands, tourism has greatly amplified the intensity and dimensions of this threat. Some community members have responded by illegally selling their land to outsiders, often fearing they will lose their land without financial compensation if they refuse to sell.[18]

While the Brazilian example illustrates some of the general threats and sources of friction that may arise from mere contact of any kind and from attempts at their integration with persons from the prevailing model of economic development, the Garífuna example highlights some of the specific threats and sources of friction arising from tourism development felt by indigenous groups.

A necessary question to ask at this stage regards the possibility that the economic activity of tourist visits to tribal groups might result in a non-destructive form of development, unlike other economic activities associated with the exploitation of natural resources. To some extent we have already addressed this question in Chapters 3 and 4, but it is relevant here because of the dependence of tribal groups on the ecosystem that supports them. Around Latin America there is a mass of examples in which proponents of ecotourism or other new forms of tourism (very often from the First World) have established tourism programmes which are sensitive and responsible to the environments and peoples that are involved as objects of tourism. Ecotourism conferences in the subcontinent often seem to be a litany of case studies of successful schemes, all pointing to the difference of this kind of activity from the earlier (and current) extractive type of economic exploitation. There can be no doubt that there are many examples of good practice in the tourism industry of the LAC countries, these including some tourist visits to tribal peoples.

It is possible, however, that these forms of tourism are just as extractive and just as exploitative as those economic activities whose object of interest is more clearly seen as tangible and physical. Tourism's object of interest is the trophy snapshot of a culture and life of such great difference from that of the exploiter that they, the tourists, must take away with them some small part of it – a photograph, a trinket, or just the memory of an unforgettable experience. But does it leave the culture and environment intact? We would venture to suggest that there is as much evidence from previous anthropological and tourism studies[19] to suggest that indigenous cultures have in many cases been adversely affected by the experience of being hosts to tourists. In fact in an earlier work[20] we characterise the experience of tribal groups as hosts by use of the fabricated word *zooification*, a term which very much speaks for itself. Essentially it captures the way in which the tourists and tour operators treat tribal groups as objects to be viewed in much the same way as 'human museum exhibits'.[21] Tribal peoples are commonly perceived as natural and wild, an intrinsic part of the nature which many tourists visit the LAC countries to see. And disappointment reigns if the visited groups show signs of adapting to western culture and economy and fail to confirm the stereotype that is expected of them. Note this disappointment in Mick Jagger when interviewed after an Amazon trip: 'What the tour guides do is take all the Indians' clothes off and put their little skirts on them, hand them a spear which they hand back at the end. The Indians dance a little around you.'[22] His comments betray a belief that his tour did not provide the expected dose of 'authenticity', the importance of which is noted by Alex Standish in his review of Jim Butcher's book *The Moralisation of Tourism*. Standish describes the New Moral Tourism as:

> seeking to preserve these cultures in their traditional and past states. In doing so it presents culture as something static and unchanging denying hosts the creative potential to advance their culture. Ultimately, New Moral Tourism seeks to turn developing world destinations into a museum for westerners who reject their own way of life, instead searching for an elusive authenticity.[23]

This authenticity of the tourist experience is brought into question by some of the examples cited in this chapter. Even in those cases where control of the visit is in the hands of the tribal people, the nature of their ceremonies or products is altered for the sake of the visiting tourists. The Kuna, for instance, carry out rehearsals of dances and songs on the night before a cruise ship is due to arrive. As the case study outlines, the Kuna are adept at staging authenticity. Knowledge of the fact that the tourist experience is staged often fails to deter the tourists from wanting to experience it – but why should it? – a reflection perhaps of the new tourist's need to collect cultural capital and kudos for their return home.

The zooification process involves turning tribal peoples into one of the 'sights' of a rainforest expedition or a trek. As Survival International has said, 'All too often tour operators treat tribal peoples as exotic objects to be enjoyed as part of the scenery'.[24] As Rigoberta Menchú, a Guatemalan Quiché Indian and Nobel Peace Prize winner, has stated, 'our costumes are considered beautiful, but it's as if the person wearing it didn't exist'.[25] On tour it is difficult to square the colour, attraction and beauty to which Rogoberta Menchú refers with William Blum's description of Guatemala's Indian population:

> It would be difficult to exaggerate the misery of the mainly Indian peasants and urban poor of Guatemala who make up three-quarters of the population of this beautiful land so favoured by American tourists . . . In a climate where everything grows, very few escape the daily ache of hunger or the progressive malnutrition . . . Highly toxic pesticides sprayed indiscriminately by airplanes, at times directly onto the heads of peasants, leave a trail of poisoning and death . . . public health services in rural areas are virtually non-existent . . . the same

for public education . . . near-total illiteracy. A few hundred families possess almost all the arable land . . . thousands of families without land, without work, jammed together in communities of cardboard and tin houses, with no running water or electricity, . . . sharing their bathing and toilet with the animal kingdom. Men on coffee plantations . . . living in circumstances closely resembling concentration camps.[26]

The process of zooifying tribal peoples inevitably leads to a position of powerlessness for them as well as a complete loss of human dignity. The key to avoiding such situations is control of and participation in the tourism activity, which does not necessarily mean simply a greater share of the financial profits. As Pretty's typology of participation[27] (see Chapter 3) suggests, it also implies control over all the conditions of the tourism development. Again, as was noted in our discussion of participation and empowerment in Chapter 3, one of the most important elements in the success of a tourism scheme is that the idea and impetus for it should come from within the community itself.

It is of course too simplistic to demand a singleminded, blanket policy of total control to the tribal groups involved in any tourist development. There are dangers, as Colchester points out, in making 'an assumption that once an area is under indigenous ownership and control the problem is solved . . . This is patently not the case.'[28] Notwithstanding these dangers, it can be argued that the community has to own and control the development if it is to avoid the pitfalls associated with external control.

The nature of the change that tribal groups experience as a result of tourism, however, may not be always or totally malign, as Pratt's notion of transculturation manages to relay to us.[29] Critics of tourism in the Third World have consistently rounded upon cultural 'bastardisation', 'trinketisation', the destruction of indigenous cultures, and so on; and certainly many indigenous groups which have either adopted tourism as a means of economic gain or have been adopted by it, participate in the interaction through the sale of trinkets. Critics contend that tourism is a process of acculturation through which Third World and Fourth World cultures are assimilated into materialistic First World lifestyles.

Pratt refers to 'contact zones', 'social spaces where disparate cultures meet, clash, and grapple with each other, often in highly asymmetrical relations of domination and subordination – like colonialism, slavery, or their aftermaths as they are lived out across the globe today'.[30] Contemporary tourism, and particularly new tourism in the Third World, is staged in these so-called contact zones, which serve to emphasise that tourism is experienced in sharply differentiated ways by visitor and host. Pratt continues: 'A "contact" perspective . . . treats the relations among colonisers and colonised, or travellers and "travelees", not in terms of separateness or apartheid, but in terms of co-presence, interaction, interlocking understandings and practices, often within radically asymmetrical relations of power.'[31]

However, this asymmetry of power between host and guest tells only half the story. First, more often than not, it is charities, social movements and tourists that talk about the rights, cultural practices and uniqueness of Third World cultures, as if tribal peoples do not have a voice (which of course many do not) and are unable to represent their own views.

Second, through a process termed transculturation, Pratt attempts to encapsulate the way in which marginalised or subordinated groups select and invent from materials transmitted to them by dominant 'metropolitan' cultures. Hall refers to this as a 'cultural strategy'[32] which operates between sharply differentiated cultures which are forced to interact. It is this process of change that those engaged in the promotion and undertaking of new and supposedly responsible forms of tourism find difficult to accept. It is a feeling that we are somehow being cheated of 'authentic experiences', that this is no longer the real thing. This search for authenticity, which might be understood as a part of the desire for (cultural) sustainability, lies at the heart of much responsible tourism activity. It is an aspect of new tourism that is sharply reflected in visits to tribal groups and their villages, an activity about contact with 'real' cultures.

One advantage of the concept of transculturation is that it allows us to explore possibilities that lie beyond the often repeated charge that tourism distorts, disrupts and bastardises Third World and Fourth World cultures. In some instances this is, of course, exactly what happens, especially where the power of the tourism industry is intense, as described in some of the case studies later in this chapter. But it is a view that debars us from considering how the visited actually adapt and borrow from cultural practices and in turn modify their own cultural practices or ways of making a living, even in circumstances where their power is differentially distributed. The youth of the Mapuche nation of Argentina and Chile, for instance, are beginning to re-assert themselves with an adapted identity which affirms their Mapuche status but in a way which reflects their contact with modernity. In 2002 they launched the Mapuche Campaign of Self-Affirmation and some refer to themselves as *mapunks* and *mapuheavy* youths through the media of drama, music and workshops. 'They are still occasionally challenged by many Mapuche activists, old and young, who insist on only one particular way of being Mapuche'[33] – one particular struggle of the Mapuche nation against the tourism industry is documented in a case study later in this chapter.

There is, then, at least a degree to which indigenous groups may participate in their own assimilation to a 'developed' state, and the case studies given later in this chapter illustrate some of these differing degrees of participation. Participation, however, has many facets as we have already explored in Chapter 3, and the empowerment that is supposed to occur as a result of participation in a scheme is not always achieved. It is worth referring each of the case studies given later in this chapter to Pretty's typology of participation given in Chapter 3, and we urge the reader to try this exercise. We also wish to refer the reader back to the discussion in Chapter 3 of participation and empowerment, issues of considerable relevance to tourism involving tribal groups.

Tourist visits to tribal groups are an important feature of new forms of tourism in the LAC countries. These new tourisms are often represented as responsible and sustainable, but, as we have already seen and as we shall see again in the case studies later in this chapter, the level of responsibility associated with tourism involving tribal groups may be very variable and can be called into question.

Photography – shooting the locals

We have positioned this brief discussion on photography in tourism in this chapter because the problems and dilemmas created by tourists' use of cameras are at their most acute where tribal groups are the targets of the camera's aim. Some tribal groups are reported to believe that their souls are removed when their image is captured by a camera. This may well be so and obviously would represent something of a trauma for those tribal peoples involved, but photographer Adrian Arbib explains that 'often it's a way of saying "we don't want our pictures taken. Go away, you interfering person."'[34]

Numerous references to photography have already been made in this chapter and earlier, and there would be little dispute that photography is perceived as an essential element of most holidays. Indeed, by many tourists it is almost seen as a right and, other than for reasons of national security or personal security, many tourists might be hard-pushed to find reasons for not taking photographs. Photographs are crystallised memories and/or trophies of the holiday. As such, they are to be collected and displayed either on the mantelpiece or the wall or to friends and family. Among the relatively wealthy new middle classes of the First World and many of the emerging international tourists of Third World countries of south-east Asia and the LAC countries, the new technology of a digital camera is becoming a common possession and enables the photograph to be emailed around the world almost as soon as it has been taken. You can share your experience with friends and family almost at the same time as having it.

The photograph is just as important an element of the new forms of tourism as it is for the traditional mass tourism experiences. The latter are less likely to involve local hosts as subjects of the photographs which tend to be aimed more at friends or inanimate objects such as buildings or landscapes. Arguably the former are more likely to involve the hosts, indigenous or not, and if indigenous the subject of the photograph is highly likely to be a member of the host community. To take a photograph of a member of a tribal group illustrates just how easy it is to consume this 'other' world, so different from home, and to collect cultural capital with which to impress. But the images of course must be authentic, and the greater the difference of the image from life at home, the greater the perceived authenticity and the greater the cultural capital to be accrued by the photographer. The desired outcome of the photograph – desired by the tourist of course – is to capture the naturalness and 'nativeness' of the Third World or Fourth World, and so the reality of the image must live up to the expectation. To see members of tribal groups in their natural setting but wearing western clothes would contradict the expected stereotype. Hence, the need for tour guides to tribal villages to warn the villagers in advance so that they may strip off their western gear and replace it with traditional garb for the sake of the visitors and their cameras. Ordinarily this is done purely for the tourists, but it can also conform to the wishes of the Indians themselves. Michael

Astor, for instance, documents the case of Kuiussi, a Suya Indian chief in Amazonia who, during a visit of journalists, warned the visitors not to take pictures of Indians wearing western clothes – 'If people see the pictures, they'll say we're not Indians, that we're mixed race – and that's not true. We are all Indians here.'[35]

That example illustrates a case where power is exercised by the hosts, but especially in the case of tours to tribal groups it is reportedly more often the case that power lies with the visitor and the tour operator. In many cases tour operators control all aspects of visits to tribal villages, instructing the villagers in what to do, what to wear and what to sell and instructing the visitors in the restrictions on what they can do. The interchange is sometimes almost choreographed. Payments to the villagers may be made in money, but may also be made in other goods such as useful items or bread rolls. The case study entitled 'Bread rolls for photos', later in this chapter, illustrates some of the issues around photographing members of tribal groups.

It also raises the question of whether permission for taking photographs should be sought by the tourist. Many tour companies now include a note about photography in their advice and information sheets for tourists. An example from South American Experience is illustrative: 'Zoom lenses are essential for capturing shy locals but please do try to ask permission',[36] and the Exodus Responsible Tourism Policy statement is stronger, including the following: 'Always ask permission before taking pictures of people, ritual events or special places like shrines. If people seem reluctant or look away then *do not* take a picture. Be careful not to cause offence through your thoughtlessness.'[37] Some of the new forms of tourism, however, such as trucking offered by companies like Dragoman, Encounter Overland and Exodus, have given rise to 'stories of tour buses travelling . . . with the windows wide open so that visitors can easily take pictures without asking permission'.[38]

In their advice to travellers, the Tearfund offers a clue as to why photography may be a sensitive issue and why permission needs to be sought: 'Don't be surprised if you're asked to pay, as this may be seen as a fair exchange'.[39] Without an exchange in return for the photograph, the nature of the relationship between the tourist and the 'host' is unequal. Even when a monetary payment is made, in the tourist's terms it is normally a pittance for a trophy that will last for the rest of their lifetime. Not only is this unequal in financial terms, but it also reflects the unequal balance of power between the two parties. The tourist has the technology and knows that it will produce a long-lasting good. For the 'host' the exchange is fleeting, ephemeral and of low value, and there is a danger that the product (the photograph) will provide a long-lasting record of their poverty and lack of dignity. This, on the other hand, is precisely what many of the new tourists want – a record that they have witnessed 'real' poverty and an 'other' world – even where the new tourism may be confused with responsibility in tourism.

It is important to note, however, that not all local hosts adopt this reluctant or begrudging attitude towards the tourists and their desire to record their visit with photographs. Some are keen to be photographed and to show off their home or themselves or their involvement in a project, and in such cases no permission is needed. For instance, Alex Hamilton describes the Kuna Indians of Panama, featured in one of the cases studies later in this chapter, 'mak[ing] a dead set at the photographers, offering not only their own images but carefully contrived little tableaux. For instance, a little girl with an umbrella sitting on a bench, smoking a pipe and affecting to launder a brightly coloured shirt.'[40] Hamilton refers to this as 'staged authenticity', but Judy Bennett, on the other hand, suggests that the Kuna themselves see the experience in a different light:

> A visitor from the Dutch Antilles described the arrival of a cruise ship in Carti as being 'like a circus', with streets lined with mola sellers, a woman pounding sugar cane, people with monkeys and birds to encourage photographs. Whilst the reaction of the visitor was that the Kuna were losing their dignity, a member of the [Kuna] Tourism Commission in Carti roared with laughter at the idea. 'It was like this when I was born. It's a way of encouraging the tourists to take photographs, so they can earn some money!'
>
> In Nalunega the local dance group performs traditional dances for the visiting tourists. There is little sign that the Kunas feel their dances are being 'commoditised', or that they are examples of 'staged authenticity': 'When we dance for the tourist we dance the traditional dances that the maestro has shown us. We don't change anything. We like to dance ANY day. If they want to watch we'll dance. If you want to watch, we'll dance!'[41]

The photograph represents a version of reality at a given time and in a given place; but it is not neutral and it is not objective. It has the power to capture a reality that the photographer wants to see, not just in contrived and posed photographs, but also in those instantly and carelessly taken snapshots. The photographer knows what he or she wants to portray to friends and family back home, and that is influenced by two important factors: first, that they were there and have been a part of that 'reality'; and second that the 'reality' shown in the photograph is so different from their 'reality' at home – and, the greater this difference, the greater the cultural capital earned by the trophy for the photographer. For the tourist photographer, the tribal peoples are often the subjects of the trophy because they represent the difference in reality, this 'otherness'. Like animals, they are there to be discovered, sighted, viewed and, ultimately, 'shot'.[42]

Case studies

The remainder of this chapter includes a range of case studies of tourism developments involving indigenous groups. Between them they illustrate a variety of issues concerning the responsibility of tourism to indigenous groups and their settlements, but they are unable to cover anything like all the issues. For each group the issues and problems differ according to their situation and, whilst the same issue may obviously affect several groups, they all react and/or relate to that issue differently. It is of course possible to make generalisations about the tourism-related issues that affect indigenous groups, and several are made in the Concluding Remarks section of this chapter, but it is important to recognise the diversity of societies, cultures and ecosystems within which indigenous groups survive and the differing ways in which they react to and modify their practices to allow for the pressures exerted upon them by tourism and all that the industry entails.

In order, the case studies are:

* The new bosses of the Rio Negro
* The militant autonomy of the Kuna of Panama
* Culture shock or profits? So much for carrying capacity.
* Andean trekking porters – modern-day slavery?
* The Mapuche and après-ski
* Bread rolls for photos
* The Toledo Ecotourism Association – a degree of success against the odds.

The new bosses of the Rio Negro

The Rio Negro is a major tributary of the Amazon in the northern Brazilian state of Amazonas. The Rio Negro basin is home to over twenty tribal groups including, among others, the Tukano, Baniwa, Baré, Piratapuia and Arapaço, Yanomami and Kuripako. Outside of the town of Barcelos, the population of the Rio Negro basin numbers about forty thousand, who live in 750 communities and in two urban centres, São Gabriel da Cachoeira and Santa Isabel do Rio Negro which have 11,000 and 4,000 inhabitants respectively. About 90 per cent of the population is indigenous and the non-indigenous population is concentrated in the two urban centres.

Since the eighteenth century, missionaries, the military and white merchants penetrated the region and influenced the tribal groups to move away from their old villages within the forest towards the river banks, along which they have spread. 'The "civilising" evangelical project implemented by the missionaries represented for the Piratapuia people a "huge cultural catastrophe"',[43] according to Rosilene Fonseca Pereira, director of the Federation of the Indigenous Organisations of the Negro River (FOIRN by its Brazilian initials). Economically, the gradual move towards the river banks made these groups more dependent on the activity of fishing than previously. Since 1994 the FOIRN and the Socioenvironmental Institute (ISA by its Brazilian initials) have worked as a partnership, one of whose achievements has been the demarcation of ten million hectares of land as indigenous territory.

During the last two decades, the local population 'have faced another wave of "white" penetration, working for national security policies or for the interests of mining companies, which brought environmental destruction and a number of cases of violence'.[44] This last wave of intruders, however, has also included tourism entrepreneurs and their clients, the tourists, the vast majority of whom arrive in the area for sport fishing advertised by a number of companies. The tourists arrive by charter plane, mostly from Miami, Florida, to Manaus, from where they are taken to their boat-hotel on the Rio Negro either by local plane or by boat. More than a dozen boat-hotels operate on the Rio Negro, without land-based tourist infrastructure, and all have small fleets of powered fishing boats. These boats are driven largely by local *técnicos* sometimes accompanied by guides who refer to the business owners of the boat-hotels as 'the new bosses of the Rio Negro'.

Two land-based hotels, the Rio Negro Lodge and the Rio Araca Lodge, along with a boat-hotel called the Amazon Queen, are owned by Phillip Marsteller, a North American, known locally as 'Felipe', who also owns Amazon Tours Inc. of Brazil. For clients of the three hotels, Amazon Tours arranges all the necessary transport within Brazil, full lodging, all meals, an open bar, six days of fishing per week, Brazilian fishing licence, all fishing equipment, guide service and daily laundry service, for a price of around US$4,000 per person per week. This price does not include the international air fare, and clearly the experience is only for the wealthy.

Amazon Tours opened the Amazon Queen floating hotel in 1992 and because of its commercial success opened the Rio Negro Lodge in 1998. The lodge was designed to:

> epitomise luxury for up to 36 guests. Spacious and air conditioned, the main lodge spans 3,000 square feet beneath a verdant jungle canopy . . . Located on paths winding through the lush rain forest grounds of the Rio Negro Lodge, the cabins feature air conditioning, queen-sized beds, private baths with hot water, walk-in showers, a sitting area and a covered porch . . . Absolute comfort . . . our guests will enjoy resort-level amenities like daily laundry service, a private beach and fine dining.[45]

More recently, in 2002, it has opened the Rio Araca Lodge for the more remote experience.

To our readers, or at least to those who enjoy the sport of fishing, this may sound a tempting experience, but not mentioned by the Amazon Tours website (perhaps not too surprisingly) are the conflicts with the local indigenous peoples which have been caused by their operations. Many of the indigenous communities in the Rio Negro basin have complained that the sport fishing motor boats scare away the fish, which make their own manual and subsistence fishing more difficult. They have also complained that 'tourists respect neither the communities nor the ranch owners, photographing and filming them without their authorisation'.[46] It is also reported that Felipe prohibited hunting by people on neighbouring ranches to his land, threatening to call the police and take away their shotguns, and pressuring some Indians to abandon their land.

In May 2002, FUNAI (the Brazilian National Foundation for Indian Affairs) suspended new constructions and tourist installations in the region pending the results of a study it was conducting into the designation of Indian territories in the region. Then in early 2003, the federal police destroyed a number of tourist encampments on the banks of the Rivers Quiunini, Aracá and Jurubaxi after ASIBA had claimed that these were on land designated as belonging to various indigenous communities.

In early 2004, the relationship between the operators of tourism businesses and the indigenous and riparian communities in the region was described as tense and worsening.[47] Along with state military police and officials from the Institute of Environmental Protection of Amazonas (IPAAM), Felipe seized fishing equipment from local fishermen in order to prohibit artesanal fishing, which is practised for subsistence, and to preserve the sport fishing that he offers to his clients. Felipe paid for the transport and accommodation of the two IPAAM officials and the military police involved in the action – 'a true association between private and public resources to bring about an end linked to the capitalisation of the exoticism of the Amazon environment'.[48] These abuses were denounced by almost 200 local fishermen, with the help of the Association of Fishermen of Barcelos and the Association of Indigenous People of Barcelos (ASIBA) who managed to get an audience with IPAAM officials, a representative of Amazon Tours Inc., and the municipal secretary of Tourism and the Environment. The result of their complaint came a few days later when federal police seized thirty of Felipe's boats and a hydro-foil on the grounds that they had been purchased outside the country and had not been correctly registered upon import. Felipe was also obliged to return the confiscated fishing equipment to the local fishermen.

Ecotourism, however, is an activity that is considered to have great potential in the area and is therefore looked upon with approval by many of the authorities. In the municipality of Barcelos, for instance, there are two ecotourism and sport fishing programmes which are supported with technical agreements between local, national and international authorities. One of these is financed by the Inter American Development Bank (IDB); the other is supported by the State Secretariat for Culture, Tourism and Sports, the Institute for Environmental Protection of Amazonas (IPAAM), the Brazilian Institute for the Environment and Natural Renewable Resources (IBAMA) and the Municipality of Barcelos. The local Indian and riparian communities claim that they were not properly consulted about either of these programmes. Given that these organisations channel funds into the development of ecotourism and sport fishing, the crucial issue concerns the destination of these funds. On the reasonable assumption that the funds go to those who wish to develop facilties for such activities and on the evidence that these are largely outside entrepreneurs, then this signifies approval of their developments. In turn, this signifies that the activities of the long-term inhabitants of the area are of lesser importance, and as Sidnei Peres describes them, 'to the detriment of the improvement of living conditions of the Indian and riparian communities'.[49]

In this case, power and capital are the all-important factors in determining who benefits from tourism.

The militant autonomy of the Kuna of Panama

The Kuna Indians inhabit the Comarca of Kuna Yala, on the San Blas Archipelago of islands off the Caribbean coast of Panama and a narrow strip of land along the mainland Caribbean coastline. In the comarca there are approximately 40,000 Kuna, although the diaspora contains another 15,000 or more living largely in the cities of Colón and Panamá.[50] The adaptation of the Kuna to the metropolitan culture – in this case it cannot be called predominant for the reasons outlined below – is such that more than a few of their youth gain scholarships for study in the USA. A majority of these return to live a kind of double life with one foot in Kuna society and the other in western urban society.

The Kuna have a high degree of autonomy which they have gained at least in part as a result of their tendency to become militant and aggressive in the face of threats to their society and culture. This tendency is partly explained by their history, which Bennett briefly summarises as follows:

> At the time of the Spanish Conquest between 300,000 and 700,000 Kuna-speaking peoples inhabited the Darién rainforest, an area that today remains one of the most inaccessible regions on earth. For 250 years the Kuna were under extreme pressure as the Spanish tried to colonise the Darién, mine the area's gold, and subject the Indians to their political control. In 1787 the Kuna emerged triumphant but with their numbers greatly reduced by warfare and disease, and entered the nineteenth century with control of large areas of the Darién region of eastern Panamá (Holloman, 1969[51]). The Kuna were almost unique in coming out of the Spanish period as an independent people, achieving this largely through alliances with English and French pirates, and through their willingness and ability to modify their social organisation. In 1831 New Spain granted them their independence (Holloman, 1969;[52] Salvador, 1976[53]).[54]

Two examples from the Kuna's twentieth-century history illustrate their tendency to militancy and its importance.

After the creation of the state of Panama (as a territory independent from Colombia) in 1903, the new Panamanian government adopted a policy of acculturation of the country's tribal groups into a culturally and economically modern national system. Traditional dress and ceremonies were banned and schools and police were introduced into tribal communities, including the San Blas islands. The general discontent with this situation led eventually to a brief revolt in 1925 in which all the police were either killed or driven off the islands. As a result, the comarca became officially autonomous in 1930, and in 1938 it gained official recognition as a Kuna reserve.[55]

Foreign ownership of land in the comarca is prohibited, and the second example of the Kuna's militancy refers to the treatment of outsiders who have tried to open tourist facilities without the required approval of the Kuna General Congress (KGC), the Kuna's governing body. Around the start of the 1970s, two North Americans, W. D. Barton and Tom Moody, opened two tourist resorts, one called 'Islandia' and the other on the island of Pidertupu. Both had gained permission of the local Kuna chiefs, but had failed to seek approval from the KGC. Both 'owners' were driven from the comarca, Barton's hotel being burned down twice and Moody being injured in a shooting shortly after the formally announced deadline for his departure had passed. For further details of both cases the reader is referred to Swain (1989)[56] and Bennett (1997).[57]

Their ability to determine their own affairs has meant that the Kuna have been the major beneficiaries of tourism developments in the comarca. But it does not mean that there have been no differences of opinion about the future direction of tourism developments; nor can it be said that the external and internal pressures resulting from tourism are insignificant. Bennett believes that these problems, differences and pressures arising from tourism are seen by the Kuna as 'contributing to an increased dependence on the cash economy with a resultant reduction in agricultural production; as a threat by outside investors to rob the Kuna of their land; and as a cause of conflict and division within the comarca'.[58]

Despite these perceptions, it would also be true to say that tourism is increasingly seen as a source of gainful employment and economic survival. Alex Hamilton (1995),[59] for instance, suggests that cruise ship passengers visiting the islands for just a couple of hours often leave around US$10,000 for all their purchases. In particular, these purchases include brightly coloured 'molas', small squares of material which can be sold separately or sewn together as towels, sheets and garments. (It is noteworthy, however, that cruise ships visit only the western end of the comarca because of the dangers of the reef further east, and this obviously limits the number of the Kuna who can derive benefit from this type of tourism.)

But their increasing dependence on tourism is a source of friction within the community. It is perhaps an oversimplification to say that in general older members of the Kuna community (although in control of some of the hotels) are keen to control and limit the influence of tourism, whilst younger members (at least those not lured to Panama City) tend to be eager to exploit the opportunities it offers and to develop it – but it nevertheless contains an element of truth. In 1996 the KGC took control of tourism policy by creating a Tourism Commission and passing a Tourism Statute, one of the major principles of which is the prevention of the loss of ownership of Kuna land or property to outside entrepreneurs through tourism development. Several illustrative articles of the statute are given in Box 5.1 and they clearly stress the point about power and control over resources and demonstrate the Kuna's awareness of the threats which tourism poses to their culture. Bennett (1998)[60] outlines some of the threats and difficulties faced by the Kuna from tourism development, despite the existence of the Statute:

> The principal form of tourism today is the cruise trade, augmented by some longer-term visitors who stay in the few hotels in San Blas. In the case of the cruise ships, local control seems to work well. But the Kuna are poor and many island communities do not benefit from the tourist income. Attitudes to visitors and to the importance of preserving the old tribal ways vary widely among the communities stretched out along the 220-kilometre chain of atolls. It is in practice hard for the KGC to police its Statute, for reasons of communications, geography and weakening ties of authority, but also in large part because the policy is highly controversial among the Kuna themselves. Unless the KGC is able to adopt a more positive, albeit still limited development of tourism, its extraordinary achievement of local control could be fatally undermined by the very communities it is intended to serve.[61]

Box 5.1 Selected articles of the Kuna Statute on tourism in Kuna Yala

Article 1 Tourist activities are only allowed in Kuna Yala as long as they pay respect to, conserve and validate the natural resources, the environment, the biodiversity, and socio-cultural, political, economic and religious values, Kuna norms and customs.

Article 7 Properties owned by the Kuna Yala community cannot be used for guarantee or to obtain a loan by any kind of tourism project. Not complying with this article may result in the cancellation of the project or lead to confiscation of goods.

Article 14 Any tourist activity has to be carried out in a way that allows continued respect towards the Kuna people, as well as the resources and the environment. Sport activities violating these principles are forbidden, such as water ski, jet-ski, micro-light planes, etc.

Article 19 It is forbidden to take photographs, tape or film the Kunas or communities without authorisation by the local authorities and the people affected.

Article 20 It is forbidden to extract or destroy any natural element whatsoever within Kuna Yala. This is also applicable for archaeological findings and any other elements.

Article 50 The right to entertain tourist activities in Kuna Yala is reserved to the Kuna only.

Article 51 Any Kuna who wishes to entertain tourist activities or set up such infrastructure needs a written authorisation by the Kuna General Congress (KGC).

Article 52 Any tourist activity that does not comply with the above articles is illegal and Congress will confiscate all goods in connection with the activity.

Within Kuna society the widespread debate around the issue of tourism development has at least led to a general awareness and widely held view that their communities cannot be treated as zooified objects for sightseeing, unless of course they control the activity themselves.

Culture shock or profits? So much for carrying capacity

In the first half of the 1990s, a private tour company named Eco-Tours SA was based in Panamá City. (This particular company no longer operates, but others may have taken its name.) It specialised in organising guided tours into Panama's Darién jungle and other pristine natural environments in Panamá and arranged visits to the settlements of indigenous groups in various regions around the country. It offered regular, scheduled tours, but also arranged bespoke tours for special parties. It also ran courses for the training of indigenous guides, and these courses included language classes in Spanish and English. In my interview with him, one of its managers spoke of a maximum group size of 16 to 20, but this number was obviously chosen arbitrarily as a carrying capacity. From the way they talked, its managers and employees seemed to be particularly aware of environmental problems, but they admitted that they had taken parties of larger than 20 in the past.

My interview with the sales manager was interrupted by a representative of a cruise ship which had just docked in Panamá City. He had approached Eco-Tours for a day trip into the jungle for some of the passengers, all of whom were wealthy Americans and virtually all of whom were over fifty years of age. They wanted the trip on the following day for the full day, and there were possibly as many as 80 passengers interested in participating. They had specifically requested to see the rainforest and unassimilated indigenous groups.

The company had five of its guides already committed to scheduled tours on the following day. That left only two qualified. But only one known and regularly used tropical rainforest route remained unused on the following day after the scheduled tours were taken, and that was not the easiest of walking tours. To put any of the cruise ship group on the same route as those already to be used on the following day would have exceeded the chosen carrying capacity.

The fact that I,[62] a supposed environmentalist, was present in the office was clearly an inhibition to their decision-making process, but a decision had to be made on the spot without regard to the sensibilities of an unimportant visitor. Perhaps not too surprisingly, the company opted to take the group despite its size and despite the fact that there were not enough guides for the company to comply with its own arbitrarily set carrying capacity. In any case, the manager promised that he would find other guides by tomorrow.

In the event, two guides took what turned out to be 70 cruise ship passengers in two groups along trails which were already in use during that day. They staggered the starting times of the scheduled groups and the cruise groups in order to avoid groups overlapping. The two tribal groups visited twice in that day received extra money to compensate them for the disturbance to their daily routines. They also made more money from the beads and trinkets which they sell to the visitors. The company itself of course charged a hefty fee, based on a per visitor rate, to the cruise ship. The passengers were reportedly delighted.

The negative effects of the company's decision in this case are much less easy to count than the money that was gained by those offering the service and by those willing to be photographed and stared at as if they were mere animals of the jungle.

The already worn trails suffered marginally greater stress than they were used to; the mainstream Panamanians involved in providing the service were reinforced in their belief that the Panamanian tribal groups were of use solely because of the weirdness factor they represent to tourists; and the growing dependence on money of the tribal groups continued. These negative effects are not given here in judgement. We are not saying that the integration of tribal groups into mainstream society is inherently wrong or unwise. We are not suggesting that tribal groups should not be introduced to and steered, gently or otherwise, towards a money economy. Some tribal groups now thrive on it. Rather, judgements are left to the reader.

Andean trekking porters – modern-day slavery?

The Andes range is home to a number of tribal groups distinct in their physical, social and cultural characteristics from Indian groups of lowland South America. The mountain range is also a major destination for mountaineers and trekking tourists. In particular, Peru, and even more particularly the ancient mountain-top city of Machu Picchu, are major attractors to those seeking the adventure, thrill and experience of altitude sickness, physical exhaustion, sunstroke and frostbite, all in one package.

The search for authenticity lies at the heart of trekking in the Andes: the authenticity offered by the rugged mountainous terrain, spectacular natural environments and indigenous local cultures. It is an activity often supposedly associated with solitude and distance from other tourists, although whether these are achieved when the majority of trekkers travel in groups accompanied by many porters is open to a little doubt. Recent years have seen many papers and articles on the negative effects of trekking, these including the loss of forest area and soil, and the streams of litter and toilet paper left by trekkers. The positive effects, such as the cash left by trekkers and the remarkable personal experiences, have also been the subject of many articles, the majority of these being the products of the trekking holidays taken by writers and academics.

Trekking parties, and to a lesser extent mountaineers too, employ porters to carry the goods and equipment required to sustain the trek. In the case of the Peruvian Andes the porters employed come largely from Andean Indian groups and it is often assumed by trekkers, trekking party organisers and tour operators that, because their kin are commonly seen carrying large bundles or baskets of goods to market and wearing little footwear, they are immune to or accustomed to the effects of physical exertion, high altitude and intense cold. In brief, the attitudes of many trekkers towards their porters is nothing short of colonial – the porters are effectively pack animals, hired to take the strain off wealthy westerners who seek an 'ultimate' experience, or what Will Hutton describes as a 'must do',[63] an experience that they must take back home with them.

The reality of the conditions endured by the porters and their ability to endure them are widely at variance with the assumptions made by trekkers:

Examples of porters working in sub-zero temperatures with huge loads, poorly dressed in flip flops or plastic bags as waterproofs have been seen time and time again. Illnesses directly related to trekking such as hypothermia and altitude sickness have consistently been reported in many situations where proper equipment and medical attention would have prevented them. Neglect of these illnesses and other accidents has even led to the death of a number of porters.[64]

The International Porter Protection Group (IPPG) was established in 1997 and, as its website states, it is 'working towards a sustainable and ethical trekking industry' and 'for the safety of the mountain porter'.[65] Their aim is 'for every porter to have access to adequate clothing, boots, shelter and food, appropriate to the altitude and weather, plus medical care when ill or injured'. Like most campaigning groups, they use a mixture of tactics to achieve these aims, including political lobbying, education, monitoring and direct action. Raising awareness of these issues amongst all players in the game of trekking tourism and mountaineering is an important aspect of their work, and the next paragraph illustrates some small measure of the group's success in achieving the cooperation of the government of Peru. But awareness of the problems and issues is still very young and thinly spread – some of the real costs of trekking are still unknown to most of those who take part in the activity.

In 2001 new regulations were introduced on the Inca Trail to Machu Picchu in order to limit the damage done to the trail. Only licensed operators are permitted to run tours along the trail and their use of porters is also regulated. For instance, their packs are now weighed by government officials at the start of each trek and a maximum of 20 kg pertaining to the agency and 5 kg for the porter's own possessions is allowed for their loads. In the same year Tourism Concern UK launched its campaign 'Trekking Wrongs: Porters' Rights' and designed a set of guidelines for UK tour operators on porters' rights and working conditions. Many of the UK operators initially contacted as a part of this campaign were unaware of the extent of the problems faced by porters and left all responsibility for their treatment and conditions to the local, contracted operators. By the end of 2003, 41 of the 80 UK tour operators contacted by Tourism Concern had developed their own guidelines and/or adopted or adapted the organisation's guidelines.

Despite the new regulations, porters continued to report poor working conditions, low pay and overweight packs with extra items added after the official weighing process. In 2002 with the help of an international campaigner (Alison Crowther), Peruvian porters themselves established their own organisation, the Inka Porter Project (Porteadores Inka Ñan). The Project achieved some success in terms of the design of new backpacks and acquisition of improved equipment. It provided trekkers with sets of suggested questions that they should ask of their trek operators, it collaborated with a range of international mountaineering and campaigning groups (such as the IPPG) and linked with a number of trekking agencies (such as Explore Worldwide and local Cusco agencies). It campaigned and lobbied on behalf of its members to raise awareness of their plight and to improve their working conditions. Despite this progress, it still had a mountain to climb in achieving its goals of improving the working and living conditions, pay, health and self-esteem of porters. This last point is as crucial as any other, as the Project's web page points out: 'The self-esteem of porters is generally extremely low and they have a very low social status. Building a more powerful self-image through participating in practical, health and environmental projects will allow them . . . to improve their conditions.'[66]

Although the Inka Porter Project's website is still active, in July 2005 the organisation ceased operations owing to a lack of staff resources. The IPPG and Tourism Concern's campaigns are still operating and active.

The Mapuche and après-ski

The Mapuche people inhabit the south of Chile and Argentina on both sides of the Andes mountains. The Mapuche was the first indigenous nation of the Americas to be recognised (in 1641) in a treaty with a European power. They vigorously defended their nation from invasion until 1885 when many were killed or forced from their homes into rural communities, town and cities. Today more than 1.5 million Mapuche live under Chilean and Argentinian rule. Their right to self-determination is denied and their lands, language and culture are under constant threat.

In terms of general development, the case of the Mapuche illustrates the dilemma faced by many indigenous groups striving for full autonomy against the efforts of the government to integrate them fully into the 'modern' and 'civilised' nation state. Many of the latter's efforts are seen by the Mapuche as 'a policy of selective repression' in pursuit of a 'process of capitalist investment in Mapuche territory; in logging, energy, tourism and other sectors', and as 'a combined manoeuvre by private entrepreneurs and government, carried out with its corresponding mass-media blitz'.[67]

The Mapuche have an ongoing battle not only against the states of Argentina and Chile to retain their autonomy over ownership of land and resources but also against the pollution caused by transnational companies. In the province of Neuquen, Argentina, for instance, they have sued the Spanish oil company Repsol for damages to their health caused by oil pollution. In the same area over recent years they have also been pitted against pollution caused by the ski resort of Chapelco.

In the first half of the twentieth century, early tourism in the area was centred on the town of Bariloche in an area often described as the 'Switzerland' of South America – like several other areas of the subcontinent – betraying the origins of the European colonisers. As Barriloche gained popularity and accessibility to 'mass' tourists, so those seeking exclusivity found it in the nearby small town of San Martin de Los Andes, where the Chapelco ski resort began to grow.

The Austral airline company and related Soltours tourism company were largely responsible for the development of the Chapelco resort, which was achieved without any form of government regulation or oversight. As a consequence of this, 'the sewage works of the winter sports centre were systematically poured into the creeks that watered the three Mapuche communities downhill'.[68] This was reported in local and national newspapers as well as by the Mapuche themselves.[69]

Located downstream of the resort are the Mapuche communities of Curruhuinca and Vera and the Puente Blanco reserve where sixty families live. Having taken water from the streams for centuries without problems, they suddenly found themselves regularly and frequently suffering from diarrhoea, urinary infections and stomach ailments. An investigation revealed that the waters were being contaminated by the discharge of effluent from the ski resort into the streams. In 1999, the Chapelco ski resort agreed to clean up its effluent discharges, but nothing resulted from this agreement.

In August 2002, a court order to close the resort was granted at the same time as the Mapuche themselves took direct action to barricade the access road to the resort. The action was quickly called off when the protestors heard of the court order, which obviated the need for counter-action against the protestors by the police who had arrived at the barricades as the action was called off. The action was timely as it coincided with an international snowboarding competition sponsored by Nokia which was attended by five thousand visitors and a sizeable contingent of the international press.

Whilst the barricade of the access road was in force, many of the visitors waited in a traffic jam 3 km long; but, instead of expressing annoyance, they sided with the protesting Mapuche after the latter had explained that if they, the tourists, entered the resort and used the bathrooms there, they, the Mapuche, would later be drinking their contributions. Later in the day, however, the court order was lifted and the tournament went ahead.

This may sound like something of what is commonly known as a stitch-up – accede to the request so that you can get your way, and then later revoke your word – but in fact the Chapelco Ski Resort has since built a sewage treatment plant which has resolved some of the water problems. In May 2005, however, Alfredo Seguel reported that, 'despite . . . the installed treatment plant, some environmental problems have still not been resolved'.[70] Also since the protest action, the resort has taken to employing snow cannons which consume much of the water that should be destined for the communities downstream.[71]

Moreover, throughout 2003, the Mapuche continued to take actions and to campaign for their autonomy over land and resources which they believe belong to them. They have continued to suffer police repression and searches and the authorities have continued to ignore agreements made with the Mapuche. Land rights settlements are still pending with respect to the Mapuche land that falls within the area of the Chapelco ski resort.[72]

Bread rolls for photos

Maria McLintock is half-Indian, half-Scottish, daughter of a Scottish geologist and a member of the Huaorani tribe in the Oriente region of Ecuador. She lives in Misahualli, a small town on the Rio Napo, which used to serve primarily as an army post and a trading post between tribal peoples, squatters and colonisers to the east and the more populous highlands of the Andes. It still serves as such, but in the last twenty years of the twentieth century the town grew in size significantly because of the arrival of tourists. Most houses in the town now offer some sort of service to the tourists.

Maria lives in the town and, like many others of its residents, offers jungle tours to the visitors as well as rooms for board and lodging. She shares the advantage of some of her fellow tour guides of being familiar since birth with the environment that she shows to the visitors. She is a mine of information about the surrounding flora and fauna and is something of a specialist in the medicinal uses to which a myriad of rainforest plants can be put. She still has links with her tribal village, although she was raised by her father and is thoroughly western in her ways.

Rising early in the morning for a full day's jungle tour with Maria were two New Zealanders, a German couple and three Brits. The pick-up truck took them, with Maria and her driver, several kilometres to the start of a short jungle route during which it seemed as if she was able to point out the name and describe the uses of all the plants under the sun. They were too numerous to remember, and the tour group members who began by taking notes soon ceased to do so as memory banks approached overflowing. By the end of the day, each member was able to recall two or three items which held some particular meaning for them. Added to the information about the wealth of vegetation were the animal droppings, tracks and other signs of wildlife such as scratch marks on tree bark. But other than ants, millipedes, butterflies and a few birds, no large wildlife was seen. Maria had to explain that, although the parcel of land they were on was lush in its vegetation, it was only one stand that was surrounded by land deforested by the oil companies and the squatters and settlers who moved in along the oil company roads. Most of the monkeys and other larger mammals have emigrated eastwards towards greater forest cover. But various pockets of forest still remain around Misahualli and occasionally the larger animals can be heard and seen there.

After this introduction, Maria and her driver transferred them all to a large dugout canoe with outboard motor for a trip downstream and further eastwards where the forest is thicker and more extensive and where they were to visit an Indian village. An hour in the boat was a rough experience for their backsides, especially through the rapids, and they were eager to stretch their legs on arriving and mooring at a flat stretch of alluvial plain where two Indians, a man and a woman both in western dress with flak jackets, were panning for gold with rudimentary equipment of a bowl, cloth and board. The tourists wanted to photograph them, and there was a special eagerness to capture a shot of the man's ears whose lobes were particularly pendulous with a hole between the lower and upper parts. The Indians, however, indicated with their hands that they wanted money for the pictures. They became a little agitated when three of the group took photographs and began to walk away without offering any payment. Along came Maria who gave both Indians several bread rolls each, which temporarily calmed their protestations. But then the other tourists began to snap away at the Indians and their panning equipment, which gave rise to further indignant protests particularly by the woman. Maria refused to give any more bread rolls to them (keeping the remainder for other members of the community to be visited later), and between Maria and the tourists there ensued a brief discussion about the merits and demerits of giving money to Indians. Maria suggested, but not particularly strongly, that there was no need to do so as they had been given bread instead. The two Germans, the two New Zealanders and one of the Brits were rather more adamant that money should not be given. Maria gently explained that the more money they are given, the more they are drawn into and become dependent on the money economy and the weaker become their ties with their traditional lifestyles. This was greeted with general nodding all round, but one of the Brits. pointed out that these two Indians were already very much a part of the money economy – they were panning for gold which they would sell to a gold merchant upstream in the town of Puyo, admittedly for very little, but this nevertheless represented their major economic activity and survival strategy. It was also mentioned that the group had taken photographs, so the transfer of money would not be in response to begging.

What exactly the two Indians thought of this strange event, a discussion about possible payment to them, taking place not ten yards away from them, we cannot be sure. They did not understand English, but they seemed to be fully aware of the issue under discussion. After several minutes most of the group walked away, following Maria, but the dissident Brit walked back to the Indians, took two more photographs and passed over to them a few sucres (the Ecuadorean currency at that time[73]) before turning and walking briskly away to join the group.

Maria led them along another jungle trail to a clearing around a sweet grapefruit tree that she obviously knew from her childhood. There they lunched and refreshed themselves as well as they could given the heat, and the dissident Brit tried to make the others feel guilty for being mean whilst they tried to make him feel guilty for being ignorant.

After the break they continued their muddy trek through the jungle for another twenty minutes before arriving at a Huaorani village composed of about twenty wooden and thatched houses raised above ground on stilts because of the frequency of flooding. The village also contained a relatively newly built schoolhouse which was used for community events. The only humans around were a few giggly and coy children dressed in western clothing and four women, also in shabby western-style dresses. One of them explained to Maria that all the men were away, either hunting, panning for gold, tending their fruit and vegetable plots, trading with others or working in the towns. Maria could do nothing other than apologise to her group of tourists – she clearly had not done her preparatory homework before the visit, but she wondered aloud to her group that absolutely no male adults were present and asked herself if this was because they were forewarned of the tourist group's arrival.

The visitors looked around the village, played with and teased the children and smiled at the women whose photographs they took. Maria left the whole large bag of bread rolls with the women, gathered her group together, gave them a short talk about the village, its residents, their way of life and their culture, and then marched them back to the boat.

The journey back upstream to Misahualli took nearly two hours against the flow, and the return was followed immediately by the washing of bodies and clothes in the Rio Napo. Despite the lack of animals and Indians, it was an experience that none of them would ever forget. The level of responsibility of their day's activity, however, probably still remains a little uncertain in all their minds.

The Toledo Ecotourism Association – a degree of success against the odds

In August 1991, in the south of Belize, the Toledo Ecotourism Association (TEA) instigated a village guesthouse project in three indigenous villages in Toledo. This was soon extended to a total of six villages, three Kekchi, two Mayan and one Garífuna. Each village built one guesthouse (to sleep eight visitors) in their own traditional style and using local materials. As extras, each guesthouse was built with a concrete floor, water tank and shared bathroom.

Visitors have their meals in the villages, each of three meals per day being provided by different homes in the village on a rotation system, with a maximum of two visitors eating in any one household. This is done in order to minimise the impact on the family, avoiding a large number of strangers in the house, provide more employment, involving more families, and to allow for a more personalised experience. Additionally, a local guide leads a maximum of three ecotourists along various nature trails to Mayan ruins and caves. Other activities in which visitors can participate include planting corn, harvesting, hunting, cooking, gathering firewood, fishing and river trips.

In each village visitors pay their money for accommodation, meals, tours and other services to one nominated person who pays each service provider a fixed fee for each service. Twenty per cent of the takings are paid to the TEA for the upkeep of their office in the main town of Punta Gorda, for publicity and co-ordination of the scheme and for promotion of the TEA's major aims of: developing alternatives to slash and burn agriculture, the major cause of rainforest destruction in southern Belize; village health and education projects; and the monitoring of a stress system to safeguard the environment and the quality of village life for the villagers and visitors alike. This means that at least 80 per cent of what the visitors pay remains in the village.

> Until our guesthouse program began, most tourists would arrive in Punta Gorda, check into a hotel, hire a guide to go to the ruins and our villages . . . our people were lucky if they could sell a bracelet, or basket. The guides from town and the visitors would sometimes tell us that we should not kill the animals or cut down the high trees because they may all die out. We sometimes wonder if they would rather we die out? Don't they know our families are hungry and we need the food?
>
> Now, . . . we can make enough income from ecotourism to buy sufficient food until we can develop profitable permanent homesite farms. This will enable us to refrain from killing these endangered animals and plants on the ecotrails and other protected, and as yet unprotected critical habitats, wetlands, watersheds, breeding grounds, nesting areas, on land, in the sea and rivers.[74]

From the start, a number of problems beset the initiative. Local businesses, hotels and lodging houses in Punta Gorda, the district capital, opposed the scheme in the belief that it would take clients away from them. Money and support from the government of Belize was slow in coming and, although both local and national suspicions of the scheme have now largely been overcome, the Belize Tourist Board and the Ministry of Tourism still tend to downplay or even omit the TEA scheme and its accomplishments from their international marketing.[75]

Initially, a homestay scheme which was already operating in the region offered competition to the TEA's guesthouse programme. Tales from the region have it that in the late 1980s a man from one of the Mayan villages, working as a barman up the coast in Placentia, suggested to tourists that they could visit his family in their home village. Some tourists stayed with his relatives and paid a fee for the privilege. The idea later spread and the TEA itself began a homestay programme before the guesthouse programme supplanted it. An Indigenous Homestay Network developed and these visits provided needed income to a few families. But villagers found that because their houses were small they often had to sleep in the same room as the visitors, this leading to a loss of privacy and often they gave up their mosquito nets for the sake of the visitors. Additionally in most involved villages, only one or a few families benefited from homestay visitors, causing envy in communities which traditionally value highly the notion of equality and the sharing of work and benefits. For the tourists, there were no norms of service and quality and so the experience varied greatly. In the early 1990s, the homestay network was asked to leave three of the villages largely because its benefits were concentrated solely on one family. The TEA guesthouse scheme is different from this in that it was set up to involve as many families as wanted to be involved in as fair a way as possible. The rotation of visitors between villages and between families within villages is an important element of the scheme. Villagers involved in the scheme underwent training in many aspects of catering for guests and in keeping records. In 1997 the TEA was awarded the *ToDo Prize* for 'Socially Responsible Ecotourism' at the world's largest tourism trade fair in Berlin, Germany.

In the mid- and late 1990s the United States Agency for International Development (USAID) presented another problem of competition to the TEA. In a misguided attempt to improve an environmentally and socially tarnished image in the region, USAID promoted the development of village-based and community-controlled tourism on condition that these developments also promoted the ideology of competition. With financing from USAID, the Belize Enterprise for Sustained Technology (BEST) built a new guesthouse in the village of Laguna which already had a TEA guesthouse. The new guesthouse caused rifts and rivalries between and within families in the village. The villages are small and, despite individual effort, enterprise and family-based cultural development, many of the customs, practices and norms of village life have depended strongly on co-operation and community action rather than on the spirit of competition. In one move, the USAID betrayed its adherence to dogmatic ideology and undermined everything that had been achieved by the TEA in its first few years of existence.

In 2001 another of the local problems, this time in the form of Hurricane Iris, demolished both the TEA and the BEST guesthouses in the village of Laguna, illustrating the indiscriminate nature of natural hazards in tropical zones but also the limited ability of poor communities to protect themselves against such hazards. Such hazards are of course outside the influence of the local community, but so too, it must seem to local people, are the effects of internationally financed development schemes such as the BEST guesthouse scheme and the Southern Highway project funded by the Inter-American Development Bank (IDB). In 1998 the IDB began funding the paving of the Southern Highway – referred to as its rehabilitation – through the south of Belize without reference to or consultation with its indigenous inhabitants and without protection for ancestral land rights. The Highway also opened up the area to logging companies, settlers and tourism resort schemes. Most of the highway south to Punta Gorda has now been rehabilitated, reducing the journey time from Belize City to Punta Gorda and thereby increasing the number of tourists visiting the area. This is goods news from the point of view of those who stand to gain financially from tourism; but the second stage of the Southern Highway project includes its extension into Guatemala, for which two routes are currently under consideration. The fear in the region is that the extension will bring not only much greater levels of trade and traffic but also greater levels of extraction of natural resources, pollution and exploitation. This fear has given rise to the formation of a large-scale development and protection plan which would provide for water-shed and environmental protection along with ownership rights for indigenous villages. The plan is referred to as the Toledo Peoples' Eco-Park Development and incorporates various tourism components. It has its origins within the indigenous communities and the hope is that it can be put in place before the Southern Highway extension to Guatemala is completed. But it has not yet attained approval from government, which will be necessary if its aims are to be achieved.

Clearly, the development of the transport infrastructure is perceived as representing a mixed blessing. For the TEA, the development or improvement of bus services along the region's side roads to the TEA villages would be of greater significance than the extension of the Southern Highway whose major beneficiaries will be international commerce and its corporations. Currently, the TEA attempts to send visitors to each village in rotation in order to share the benefits of the scheme, but some of the more remote villages such as San José and Barranco are difficult to reach by public transport. (During the rainy season, Barranco is accessible only by boat.) Villages closer to Punta Gorda such as Laguna tend to host more visits than others, because visitors are captive to an intermittent public transport system or have to hire their own vehicle which is an expensive option. Although ownership of a vehicle to transport visitors to the TEA villages would be useful, the TEA has so far ruled out the idea on the grounds that they need support for their scheme from local businesses such as taxi drivers. Makario Salam, a Mopan Indian involved in the scheme, suggests that it is important to show local businesses that the scheme has an economic spin-off for them.[76]

The TEA continues to survive despite its problems and is an example of a small-scale, local tourism initiative that has managed from a position of relative powerlessness to surmount local, national and international forces pitted against it. Moreover, it has done so whilst always bearing in mind a range of objectives associated with social and environmental issues rather than forging ahead single mindedly in pursuit of financial gain.

Concluding remarks

The conflicts between tourism and local cultures and peoples are nowhere sharper than in the case of visits to indigenous groups and their settlements. Responsibility for the effects of tourism on indigenous groups might be considered to lie principally with the tourists and their tour agents. But the issues of responsibility are not always so cut-and-dried. Many indigenous cultures may have remained the same with few changes for a long time in historical terms, but these cultures are no less subject to modification through dynamic processes than are other, more modernised indigenous cultures. Consideration of Pratt's notion of transculturation shows us that indigenous groups select from a predominant culture those features or characteristics (such as baseball caps, televisions and mobile phones) that they like or accept, rejecting others, and adapt their own practices to incorporate new and invading practices and ideas. Clearly, though, some, indeed many, indigenous groups have been overwhelmed and even extinguished by the intrusion of modern, western cultures, and the values associated with them. The work of Survival International has amply demonstrated such drastic effects as well as the fact that tourism has been one of the vehicles by which these intrusions have been made. But in a few cases tourism also represents a means by which some indigenous groups can enter the money economy without the pollution and threats associated with other industries. It is not an easy or clear dividing line between opportunity and danger, and each visitor has to balance the potential of such visits for doing harm against the potential for bringing benefit. Equally, each indigenous group has to weigh up the dangers against the benefits – an exercise of power which they are not often in a position to undertake. 'Although many non-governmental organisations, governments, environmental/scientific groups, and private initiatives are working with indigenous communities to develop sustainable tourism or ecotourism, it is debatable whether indigenous peoples are in a position of true equality in these relationships.'[77]

The case studies used in this chapter have illustrated a range of problems and conflicts associated with this field of tourism. By way of summary we offer the list of problems below, but caution that the list is far from all-inclusive.

- There is no simple rule-of-thumb regarding the interaction between the tourism industry and indigenous groups. Likewise, there is no simple or single way in which indigenous groups react to their introduction to the money economy.
- Attempts by national governments to assimilate indigenous groups into a 'national' identity normally provoke a strengthening of indigenous identity, which may serve as an attractor to tourists.
- Unfair exploitation and mis-representation of indigenous groups by the tourism

industry will occur wherever the indigenous groups themselves do not have control of the activities of the industry.

- Such exploitation and mis-representation includes the portrayal of indigenous groups in an environmentally deterministic way – in other words, as if they were members of the animal species, which humans are, but not of the human species. This may be referred to as the 'zooification' of indigenous groups.
- The tourist–tribal interaction is virtually always a falsely contrived one which puts the tourist in the role of voyeur and the tribal member in the role of zooified object.
- Notwithstanding the above, such interaction can be employed to the advantage of indigenous groups as well as to the advantage of the tourist.

Notes

1 Hall, S. (1995) 'New Cultures for Old', in D. Massey and P. Jess (eds) *A Place in the World? Places, Cultures and Globalisation*, Milton Keynes: Open University Press.

2 Eduardo Galeano (1973) *Open Veins of Latin America: Five Centuries of the Pillage of a Continent*, New York and London: Monthly Review Press, p. 50.

3 Progressio (formerly the Catholic Institute for International Relations, CIIR) (undated) *Brazil: Democracy and Development*, Comment Series, London: CIIR, p. 25.

4 Survival International is a worldwide organisation supporting tribal peoples. It stands for their right to decide their own future and helps them protect their lives, lands and human rights. Its publicity material explains the premise on which its work is based: 'Tribal peoples around the world are being murdered, persecuted and robbed of their lands. Oppressive governments, voracious multinationals and banks are all part of a culture of injustice and intolerance.' And of tourism it states: 'Tourism need not be a destructive force for tribal peoples but unfortunately it usually is: any tourism which violates tribal peoples' rights should be opposed. Tourism must be subject to the decisions made by tribal peoples themselves.'

5 Survival International (1992) '500 Years on – and the European Invasion Continues', *Survival International* newsletter, no. 30: 9.

6 Marcus Colchester (ed.) (1985) *An End to Laughter? Tribal Peoples and Economic Development*, London: Survival International, p. 5.

7 Chris Jochnick (2001) 'Perilous Prosperity', *New Internationalist* 335, June: 20–1.

8 Bianca Jagger (2003) Speech by Bianca Jagger in Quito Following Her Tour of Amazon Region Contaminated by ChevronTexaco, 11 October, on www. amazonwatch.org.

9 Reported for instance by Erwin Kräutler (1990) *Indians and Ecology in Brazil*, CIIR Justice Paper No. 13, London: Catholic Institute for International Relations. Also as reported by Bruce Albert and Marcus Colchester (1985) 'Brazil: Recent Developments in the Situation of the Yanomami', in Survival International, *An End to Laughter? Tribal Peoples and Economic Development*, London: Survival International.

10 Report of the CIMI (Conselho Indigenista Missionario), dated 25 February 1999, noted by Glenn Switkes (Director of the Latin America Programme of the International Rivers Network) in 'Brazil: concern over new President of FUNAI' on www.amazonia.net.

11 Joelle Diderich (1999) 'Brazil Tells Indians "Show Me The Money"', Reuters Limited, 10 June, http://forests.org/archive/brazil/showmemo.htm.

12 Instituto Socioambiental (2002) 'Indigenous Park of the Xingu', Instituto Socio-ambiental, December, www.socioambiental.org/website/pib/epienglish/xingu.

13 Jan Rocha (2002) 'Obituary: Orlando Villas Boas', London: *Guardian Weekly*, 19–25 December.

14 *Ibid.*
15 Michael Astor (2004) 'Brazil Tribes Fear for Their Environment', Environmental News Network website, www.enn.com/news/2004-02-24/s_13396.asp.
16 Louise Rimmer (2004) 'A Blood-stained Struggle for Ancestral Lands', in *Stories and Lives: 21st Century Tribal Peoples*, London: Survival International.
17 Mairawe Kayabi, quoted in Michael Astor (2004).
18 Eva T. Thorne (2004) 'Land Rights and Garífuna Identity', in *NACLA Report on the Americas* 38, 2, September/October, New York: NACLA.
19 See for instance Valene Smith (1989) *Hosts and Guests*, Philadelphia: University of Pennsylvania Press; Krippendorf, J. (1987) *The Holiday-makers: Understanding the Impact of Leisure and Travel*, London: Heinemann; De Kadt, E. (ed.) *Tourism: Passport to Development?*, Oxford: Oxford University Press; many issues of *Cultural Survival Quarterly*; a number of issues of the newsletters and reports of Survival International, the Tourism Investigation and Monitoring Team, Tourism Concern; Colchester, M. (1994) 'Salvaging Nature: Indigenous Peoples, Protected Areas and Biodiversity Conservation', discussion paper, Geneva: United Nations Research Institute for Social Development; and a host of other reports.
20 Martin Mowforth and Ian Munt (2003) *Tourism and Sustainability: Development and New Tourism in the Third World*, Second edition, London: Routledge, pp. 246–8.
21 World Council of Churches (2002) 'Statement of Indigenous Peoples Interfaith Dialogue on Globalisation and Tourism', 14–18 January, Chiang Rai, Thailand.
22 Mick Jagger quoted in Survival International (1991) 'Tourism: Special Issue', *Survival* 28.
23 Alex Standish (2004) 'The New Moral Tourism', TIM-team Clearinghouse, Bangkok, Thailand, September. A review of Jim Butcher (2003) *The Moralisation of Tourism: Sun, Sand and . . . Saving the World*, London: Routledge.
24 Survival International (1995) 'Tourism and Tribal Peoples', background sheet, London: Survival International.
25 Rigoberta Menchú, quoted in Survival International (1995).
26 William Blum (2003) *Killing Hope: US Military and CIA Interventions since World War II*, London: Zed Books, p. 229.
27 See Pretty, J. (1995) 'The Many Interpretations of Participation', *In Focus* 16: 4–5; and Pretty, J. and Hine, R. (1999) *Participatory Appraisal for Community Assessment: Principles and Methods*, Centre for Environment and Society, University of Essex.
28 Marcus Colchester (1994), p. 57.
29 Pratt, M. (1992) *Imperial Eyes: Travel Writing and Transculturation*, London: Routledge.
30 *Ibid.*, p. 4.
31 *Ibid.*, p. 7.
32 Hall, S. (1995) 'New Cultures for Old'.
33 NACLA Report on the America (2006) 'Mapuche, Mapunky, Mapuheavy', New York: *NACLA Report on the Americas*, 39: 6, May/June: 48.
34 Adrian Arbib in conversation with Sue Wheat (2001) 'Picture Perfect', *In Focus*, 39, London: Tourism Concern: 10–11.
35 Michael Astor (2004).
36 South American Experience (2004) 'Useful Information While You Are Away', www.southamericanexperience.co.uk/usefulinformation/while_away.htm.
37 Exodus (2004) Exodus Responsible Tourism Policy, www.exodus.co.uk/restourism.html.
38 Elspeth Young (1995) *Third World in the First World: Development and Indigenous Peoples*, London: Routledge, p. 220.
39 Tearfund (2000) 'A Tearfund Guide to Tourism: Don't Forget Your Ethics', Second edition, Teddington: Tearfund, p. 19.

40 Hamilton, A. (1995) 'Ecocentrics', London: *The Guardian*, 17 June: 54–6, 59.
41 Judy Bennett (1998a) 'If You Want to Watch, We'll Dance!', *Tourism in Focus*, 29: 12, Autumn: Tourism Concern.
42 Susan Sontag (1979) *On Photography*, London: Penguin.
43 Instituto Socioambiental website (2004) www.socioambiental.org.br.
44 *Ibid*.
45 Amazon Tours Inc. website (2004) www.peacockbassfishing.com.
46 Sidnei Peres (2000) '"Ecotourism" Harms Indian and Riparian Communities in Barcelos', ISA website, www.socioambiental.org.br.
47 Sidnei Peres, Professor of Anthropology, Universidade Federal Fluminense, personal communication, 15 April 2004.
48 Sidnei Peres, personal communication, 15 April 2004.
49 Sidnei Peres (2000).
50 Judy Bennett (1997) 'San Blas: The Role of Control and Community Participation in Sustainable Tourism Development', University of North London, MA Dissertation.
51 Holloman, R. (1969) 'Developmental Change in San Blas', Evanston, Illinois: Northwestern University, PhD thesis.
52 *Ibid*.
53 Salvador, M. L. (1976) 'The Clothing Arts of the Cuna of San Blas, Panama', in Graburn, N. H. H. (ed.) *Ethnic and Tourist Arts: Cultural Expressions from the Fourth World*, Berkeley and Los Angeles and London: University of California Press, pp. 165–82.
54 Judy Bennett (1997), p. 9.
55 Howe, J. (1986) *The Kuna Gathering: Contemporary Village Politics in Panama*, Latin American Monographs, No. 67, Austin: University of Texas Press.
56 Swain, M. B. (1989) 'Gender Roles in Indigenous Tourism: Kuna Mola, Kuna Yala, and Cultural Survival', in Smith, V. L. (ed.) *Hosts and Guests: The Anthropology of Tourism*, Philadelphia: University of Pennsylvania Press.
57 Judy Bennett (1997).
58 Judy Bennett (1997), p. 23.
59 Hamilton, A. (1995) 'Ecocentrics', *The Guardian*, 17 June: 54–6, 59.
60 Judy Bennett (1998b) 'Controlling Tourism through Statutes: Does It Work?', *Tourism in Focus*, 29: 11–13, Autumn: Tourism Concern.
61 Judy Bennett (1998b), p. 11.
62 Martin Mowforth.
63 Will Hutton (2003) 'Walking on Clouds in the Footsteps of the Incas', *The Observer*, 12 October, London.
64 Tourism Concern (2003) 'Tourism and Human Rights Campaign', *Annual Report*, 2003; London: Tourism Concern.
65 International Porter Protection Group (IPPG) website (2004) www.ippg.net.
66 Inka Porter Project website (2004) www.peruweb.org/porters/.
67 Coordinadora Mapuche Arauco-Malleco (1999) Public Statement, in http://webs.demasiado.com/arauko_malleko/comu6ingles.htm.
68 Nestor Miguel Gorojovsky (2002) 'Patagonia, Mapuche *et al*.', in Marxism mailing list: http://archives.econ.utah.edu/archives/marxism/2002w35/msg00144.htm.
69 Liliana Rodríguez (2002) 'A la producción del programa televesivo Punto Doc', 31 August 2002, Mapuche International Link website, www.mapuche-nation.org.
70 Alfredo Seguel (2005) Mapuexpress – Informativo Mapuche website www.mapuexpress.net, 9 May.
71 Angel Accinelli (2004) Personal communication, 28 June 2004. Angel Accinelli is the technical representative of the Mapuche Confederation of Neuquen.
72 Alfredo Seguel (2005).
73 Ecuador has since adopted the US dollar as its currency.

74 William Schmidt (ed.) (1991) *Toledo Ecotourism Association Report*, November 1990 – November 1991, Punta Gorda, Belize.
75 Toledo Ecotourism Association Newsletter (2000) 'Economic Discrimination Against the TEA?', Punta Gorda, Belize: TEA, pp. 5–6.
76 Makario Salam (2004) Personal communication, Punta Gorda, 11 August 2004.
77 Deborah McLaren, Roy Taylor and Dave Lacey (1998) 'Your Tourist Attraction, Our Way of Life', *Tourism in Focus*, Tourism Concern: London, Autumn, p. 4.

6   Urban tourism: the heart of darkness?

Introduction

Why is it that viewed from the 'West' there is a certain nobility, sympathy, even romanticism, of the rural poor and poverty, but a neglect – close to despite – for the poor that crowd into the ramshackled slums that cling precariously to the margins of Latin American cities? A strange question perhaps in a book on tourism. However, there has always been a tendency in the alternative and responsible travel world to accentuate the possibilities and potentials of culture-rich rural repositories and to downgrade or worse demonise cities in the horror stories that cram the broadsheet alternative travel pages and the 'thumpers' of travel writing. Moreover, rural areas are those demanded by the new travel, whilst the obesity of urbanisation is anything but 'cultural'. Why are First World cities, the crucibles of innovations, cultural hybridity and excitement, celebrated in the West and, comparatively, those of the developing world cursed?

Paul Theroux's sense of hopelessness is perhaps emblematic of the felt experience when we gaze as travellers on this chaotic *heart of darkness*:

> These huts, in a horrific slum outside San Salvador, are the worst I saw in Latin America. Rural poverty is bad, but there is hope in a pumpkin field, or the sight of chickens, or a field of cattle which, even if they are not owned by the people in the huts, offer opportunities to the hungry cattle rustler. But this slum outside Guatemala City, a derangement of feeble huts made out of paper and tin, was as hopeless as any I had ever seen in my life. The people who lived here, I found out, were those who had been made homeless in the last earthquake – refugees who had been here for two years and would probably stay until they died, or until the government dispersed them, and set fire to the shacks, so that tourists would not be upset by this dismal sight. The huts were made out of waste lumber and tree branches, cardboard and bits of plastic, rags, car doors and palm fronds, metal signboards that had been abstracted from poles, and grass woven into chicken-wire. And the slum, which remained in view for twenty minutes – miles of it – smouldered; near each house was a small cooking fire, with a blackened tin can simmering on it. Children rise early in the tropics; this seemed to be an entire slum of children, very dirty ones, with their noses running, waving at the train from curtains of yellow fog.[1]

So is there an alternative way of seeing and travelling in cities that charts a route between this unmitigating hopelessness on the one hand, and the colonial heritage and slick city centre bars on the other? And are there alternative ways of interpreting and reading many of the city landscapes experienced by travellers? In a counter-intuitive twist a renowned city commentator, Babar Mumtaz, suggests there is. 'In my opinion,' he argues, 'slums are not only inevitable, they are a mark of success of a city.' 'Insisting on a "city without slums", especially when no alternative housing has been developed, can mean even more hardship for the very group that is so essential to urban development: the rural migrant.' Mumtaz concludes, 'Just as slums and slum dwellers need cities to survive, so do cities need slums to thrive'.[2]

But what of cities anyway – why the fuss? Tourism – travelling – is quintessentially urban. Despite the urge to come and go, to 'get out of there' with great haste, there is little one can do to avoid the centrality that cities and towns perform in the movement and accommodation of tourists, nor the potential economic impact that urban-based tourism can contribute. Thus, as Colantonio and Potter observe, tourism became central to Cuba's re-entry into the global economy in the 1990s and 'urban tourism in Havana has been pivotal' in this political and economic process.[3] In the years 1995 to 1998 total revenue from tourism in Havana was a little over US$1410 million (48 per cent of the national total tourism income) with 54 per cent of all visitors to Cuba visiting Havana in 1999.

Beyond this, the cities and towns that make up Latin America and the Caribbean are a palimpsest of the cultural and ethnic mix that makes the modern-day region, forged from a long history of in-migration. Like cities anywhere they tell a story about the political and economic integration and relationships between the states that make up the Americas.

Arguably of more significance, if alternative tourism has something to offer the region beyond individual self-fulfilment and gratification, it is in an ability to be part and parcel of demonstrating positive change rather than a less damaging form of more conventional approaches to taking holidays. In this chapter, there is a more tangible link between tourism and so-called 'development' and in part this is expressed by our role as tourists in the increasing significance of the region's cities. The fact is that we live in an increasingly urbanised world, and at least in our lifetime this is likely to accelerate rather than reverse. A recent report by the United Nation's 'city agency', UN-HABITAT, confirms that, like it or not, we are only half-way through the global urban transition, and that by 2001 close to a third of all four hundred million Latin America and Caribbean urban dwellers lived in 'slums', and the region's urban population ran at three-quarters of the total population.[4] And of the so-called Third World regions, Latin America was already the most urbanised by the mid-1990s.[5] The challenge to development, however, lies not in the numbers or size of cities but in the socio-economic reality. As Angotti summarises, the 'Latin American metropolis is characterized by mass poverty and severe environmental pollution unparalleled in the North'.[6]

To unpack the potential of cities we need to start with exorcising the demons and consider why cities are generally vilified within the field of development.

Stinking cities: urbanisation as anti-development

Having chugged his way through Latin America in the memorable travel account *The Old Patagonian Express*, Paul Theroux's assessment of many of the Latin American city landscapes is not encouraging. 'I had no strong desire to see Mexico City again', says Theroux, 'It is, supremely, a place for getting lost in, a smog-plagued metropolis of mammoth proportions.' Theroux is no less gracious towards cities further south in Central America. Guatemala City is brutal and on its back where 'east of the capital, on the other side of the tracks . . . desolation lies . . . For a full hour as the train moves there is nothing but stone-age horror of litter huts.'[7]

If responsible and alternative tourism is, in part, an attempt to get into those cracks and crannies of development, to those marginalised by the mainstream mass tourism product which concentrates cash and power in the hands of big companies, then it is indeed an attempt to support those who have least (if any) economic clout. But tourism like development suffers from the same blind spot. Twenty years past, cities were barely on the radar screen of so-called development agencies. Today they are there, but the signal is weak, if indistinguishable. Why? For the purposes of tourism they are dirty, unhealthy, violent, noisy, alienating – and are the stuff of constructing hair-curling travelling tales.

> Someone produced a bottle of whisky, and the stories began . . . It is a custom among travellers in South America to put at least two hours a day for telling stories. They are rarely pleasant . . . A Canadian begins. I heard this in Belem, about an English guy, first day in Brazil, clean off the plane from London, and he was on this bus looking for somewhere to stay when four guys jumped him by the turnstile. Pinned him down on the ground – and get this – they rammed a fork in his arse and cut his pack off his back and threw it out the window to another guy. Jeez, can you believe it? A fork – they robbed him with a fork. More whistles of admiration. Even I liked that story, and was very impressed with the way it had remained so accurate as it travelled 6,000 kilometres up the coast from Rio . . . You see, it was my story and I still have the fork marks to prove it . . . Finally, a note of encouragement. Just before I flew home from Colombia I was in a bus queue. The bus arrived and the queue surged forward. I put both hands in my pockets – it becomes second nature – and waited for someone to try something. Then I noticed a loss of weight from the shoulder around which my camera had been hanging, hiding on the inside of my jumper. Wheeling round I caught the first throat which came to hand. It belonged to a young lad; he was holding my camera in one hand and pair of wire snips in the other. I pulled my knee up in his crotch and he gasped, dropping the camera into my outstretched hand. The whole queue broke into applause. It was a sweet moment.[8]

And at the backend of wars that gripped Central America in the 1980s and early 1990s and resulted in the deaths of tens of thousands, the best the 'intelligent' alternative travel pages could offer us was titillation of misery and suffering:

'How is San Salvador these days?' we asked the driver innocently, aware that most taxi drivers in El Salvador moonlight as paid informers. 'Muy pacífico,' came the confident reply. But every time he stopped at a red light, his head would turn slowly to the left and then the right, and his steady gaze was invariably returned . . . Emerging into the bright sunshine, we found a young man face down in the gutter. We thought he was dead, in El Salvador a fair assumption. But he was merely dead drunk.[9]

Clearly, crime and insecurity are significant barriers to both development and the expansion of tourism. Latin America saw a dramatic increase in crime and violence in the last quarter of the twentieth century, and crime is acknowledged as a serious social and economic problem especially in urban areas where it is at an unprece-dented level.[10] 'Rapid urbanization, persistent poverty and inequality, political violence, the more organized nature of crime, and the emergence of illegal drug use and drug trafficking' are commonly referenced as major causes for such increases.[11] As might be expected, the economic impacts of crime, including the tourism industry, are equally marked. For example, Guerrero estimated that the financial costs of murders in Latin America ran above US$27,000 million each year and that 14 per cent of the region's GDP was lost to violence.[12] The root causes of violence appear to be found in political, economic and social factors.[13] For Latin America, many countries have experienced protracted periods of political violence, civil wars, unrest and repressive authoritarian regimes. As Winton suggests, there are 'major repercussions of sustained state repression'[14] and many newly democratic states are yet to reform the judicial system and state policing with the result that there has been 'no systematic dismantling of past institutional structures of terror and oppression'.[15] But such violence is not confined only to so-called post-conflict transitional states. Duncan and Woolcock[16] note that poor urban communities in Kingston (Jamaica) are impacted by strict political clientelism arising from the two dominant national parties. Most significantly, as Caroline Moser reviews, the 'sheer scale of violence in the poor areas or slums means that, in many contexts, it has become "routinized" or "normalized" into the functional reality of daily life'.[17]

If the demonisation of cities were the reserve of such 'opinion-formers' introduced above, there would, arguably, be less to worry about. But there is a far more deep-seated prejudice against cities that runs to the heart of the meaning of development and underlies the politics of tourism in cities. As the UK-based charity WaterAid ponder:

the vilification of cities and their marginalisation within development . . . is not just reserved to critical academic debate, it has a clear resonance in the practice of development too . . . cities have struggled to get on the agendas of multilateral, bilateral and the burgeoning international non-government sector, and the few INGOs that have addressed urbanisation have had to initially dismantle the 'demonisation' of cities.[18]

Many reasons have been offered for why cities have been dumped in development,[19] not least that despite the accelerating growth in urban poverty it 'gets ignored because it happens slowly, inexorably'.[20] One of the most persistent problems has been the dominance of the 'rural' in development and the casting of cities as a significant part of the development problem by encouraging people to migrate from the land to the 'stinking holes'. Indeed, for some countries, the challenge has been interpreted as the need to manufacture legislative restrictions on the movement of poor rural migrants to urban areas. As Angotti argues, favoured urban strategies throughout Latin America 'often target cities themselves as the problem, and seek to stop urban growth instead of improving the urban – and rural – quality of life'.[21] Neither does it help that there is a sizeable deficit in recording and promoting the potential of cities. As part of the UK's submission to the Fourth World Urban Forum in 2006 reflects, 'people need positive and progressive examples in order to develop visions and make demands of their leaders, but virtually all of the current press coverage and commentary on urban growth and change is negative, alarmist and doom-laden'.[22]

As Deborah Eade of the international development charity Oxfam argues, the reluctance to engage in urbanisation issues is further exacerbated by the international non-government sectors' virtual silence. If cities make the headlines you can be sure it's on the back of a disaster – and as critical as this is, it is an attitude that simply strolls past the lives of millions of city dwellers each day. Eade's conclusion is:

> The prevailing attitude is either that cities are a problem in and of themselves and shouldn't be encouraged, or that their residents enjoy better facilities and so are less 'needy' than their rural counterparts, or that the challenges posed by rapid urbanisation are simply too big, too expensive, and too complicated to handle. A glance through the grants lists and literature of some of the best-known international NGOs suggests that . . . if they get involved at all, most find it easier to deal with the specific problems of specific population groups in the towns and cities of the South – street-children and sex-workers topping the list – rather than getting involved in the messier processes of urban management . . . Ironically, the largest human settlements in which many NGOs take a more holistic approach to the planning and management of basic services are refugee camps – usually cramped and often squalid settlements that earn their description as 'rural slums'.[23]

There are compelling reasons for looking at a critical discussion of cities and tourism in development, that bundle of ideas and analysis in the practice of the development industry. And it involves looking in the mirror. There is a significant heritage to development, and alternative tourism too, that argues that it is the industrial, metropolitan core of the so-called 'developed' world that maps, translates and dictates the development process, and that cities play a significant and (for some) increasingly powerful role within this global (urban) infrastructure. Cities rule the world! But at the same time development has cultivated a pedigree that leads us to the conclusion that development is, or should be, everything that the 'metropolitan world', and by inference, urbanisation is not.[24]

This 'ruralisation' of development as a core strategy asserted its prominence in the influential and popular environmental shockers of the late 1960s and 1970s. Schumacher's *Small Is Beautiful* (1974) for example effectively showed the door to cities and was part of an intellectual movement that added up to an anti-urban lobby that vilified cities within development.[25] Schumacher rounded on the 'foot-looseness' which led to urban edema and sucks the vitality out of rural areas:

> let me take the case of Peru. The capital city, Lima, . . . had a population of 175,000 in the early 1920s, just fifty years ago. Its population is now approaching three million. The once beautiful Spanish city is now infested by slums, surrounded by misery-belts that are crawling up the Andes. But this is not all. People are arriving from rural areas at the rate of a thousand a day – and nobody knows what to do with them . . . Nobody knows how to stop the drift.[26]

So, cities have been marginalised in part through their close association with industrial modernity, and in particular through their interrelationship with economic growth, environmental degradation, both locally and globally, and the often squalid living conditions and social problems experienced by poor city dwellers. Development has been interpreted therefore as ostensibly about addressing rural change, supporting rural livelihoods and stemming urban migration.

Cities as economic machines

There is a strange irony that the Fifth World Trade Organisation (WTO/OMC) Ministerial Conference (10–14 September 2003), and the ensuing protests that follow these global talk fests, were held in Cancún, Mexico. Cancún – a mini Manhattan-on-sea for sun seeking tourists – is both a concrete example and a symbol of economic conquest over nature and society. For any visitor – friend or foe – it is an amazing and frightening experiment. As environmental journalist Joel Simons explains, Cancún was an experiment to create a tourist city in paradise, a snub to the urban decay and 'third world chaos' inflicting Acapulco and chasing tourists away to competing destinations.[27] How a fishing village of 800 has transformed itself into a seaside city attracting a transient tourist population of one and half million annually is history.

With the door slammed shut on Cuba's lucrative tourism industry following the 1959 Revolution, and with the US embargo against travel to Castro's Cuba, the East Coast market was open to takers. Cancún, just one hour's flying time from Havana and less than two hours from Miami, was Mexico's response to soaking up the demand and capturing the tourist dollar. But as with experiments of creating cities by the sea elsewhere (globally and in Latin America and the Caribbean), Cancún flatters to deceive. The tourist strip of mass tourist hotels is in stark contrast to the thousands that live without adequate water and sanitation systems. As with all land and urban development, the stakes are high and the outcome of such experiments sadly predictable in their balance of power. As Simon concludes:

> megaprojects like Cancún may have made money for the federal government, they have also spawned a large-scale, capital intensive style of development geared toward quick returns. They have institutionalized land speculation and fomented a tourist economy in which capital is highly concentrated, developers are used to thinking big, and power is concentrated in the hands of the federal bureaucracy.[28]

Whilst conceived well before the full force of neoliberalism blew in the direction of Latin America and the Caribbean, Cancún is a child of what has been termed a 'lost decade for development'.[29] The 1980s were symbolised by a Reagan–Thatcher axis bent on releasing the power of the 'free market'. And cities represented a dog's bowl as a form of economic fundamentalism emerged characterised by a belief in the dominance and application of free market principles and trickle-down economic growth strategies – see Chapters 2 and 3. Trickle-down urban regeneration was relentlessly pursued in western cities, and economic structural adjustment policies in the Third World had marked impacts upon cities through the drive to implement the so-called neoliberal 'Washington Consensus' – see Chapter 2. By the close of the decade the pattern of urban development assisted by the policy consensus was characterised by 'high levels of spatial segregation; uneven distribution of population densities, infrastructure and services; large areas of under-utilised space and facilities; and a marked and escalating deterioration in environmental and social living conditions'.[30]

Hence neoliberalism has had a marked impact on the cities and towns of Latin America and the Caribbean. The principle was straightforward: privatise, liberalise, deregulate. Allow the 'magic of the market' to dictate and manage urban growth and development. On the one hand, the product is significantly changed spatial patterns and cities that construct and maintain a viable and attractive tourist product and push the problems, such as marginalised and unsightly urban dwellers cramming into urban areas, into the so-called peri-urban fringe – slums in other words. What are left are the pockets of heritage and refined culture that are necessary for tourism to survive. But not all is gloomy as the slum areas are often the zenith of innovation, entrepreneuralism and survival. Cities need to be economically viable – that's not in question. Rather it is the price that is paid and strategies employed that sweep aside the 'problems' (in some cases murderously[31]) and boost the economy in favour of the already privileged and wealthy at the cost of the already excluded and poor.

Although the fundamentalism of economic growth is now widely challenged, and the primacy of economics tempered by the need to demonstrably improve the lot of the (urban) poor, economic fanatics remain and the shadow of the primacy of economic growth in development remains long. There is a critical problem in defusing the correlation that cities and urbanisation are primarily (and in some cases exclusively) an economic phenomenon in even the most pro-poor, anti-poverty, inspired tracts. As Sachs concludes for example,

Modern economic growth is accompanied first and foremost by *urbanization* characterised by the advantage of high-density urban life . . . Once the labour force is no longer engaged mainly in food production, it is natural that the bulk of the population relocate to cities, drawn by higher wages that in turn reflect the higher productivity of work in densely settled urban areas.[32]

The World Bank's position is of special interest in three respects: first, as the largest provider of urban development assistance; second, in the influence of its policy on other lenders, agency policy and national government interventions; and third, for the significance of the World Bank in initiating, hosting and galvanising support for new global institutions, mechanisms and the harmonisation of development thinking on cities. The World Bank's renewed *Urban and Local Government Strategy* (2000) refocused multilateral attention back on to cities in the face of rapid urbanisation with the central strategic aim of promoting sustainable human settlements as defined by four characteristics: liveability, competitiveness, good governance and management, and bankability. Although the strategy emphasises, and is preconditioned by, the significance of more highly integrated and holistic interventions than previously, and comprehensive development frameworks for the 'urban arena',[33] it is nonetheless steeped in the discourse and primacy of economic health and competitiveness. Indeed the focus on cities is in part driven by the agglomeration economies yielded by economic spatial proximity not as a social or environmental entity but as an 'urban economic area that represents an integral market', with the 'prospects for economic development' and requirements for 'enhancing city productivity'.[34]

Translating this rationalised urban development (and urban form) to serve the interests of tourism as an economic sector is predictable. The World Bank concludes that:

> Tourism is becoming important in many cities, both for end-visits and as transit points in the transport system . . . Tourism, however, requires a well-ordered city, secure, clean, and healthy; that is to say, the quality of life in the city is fundamental to its capacity to earn income from tourism.[35]

These are views underpinned by an economic philosophy of 'competitiveness' and 'bankability', an approach that necessitates adopting a range of economic measures and incentives including a commercial approach to many urban services and functions whilst 'keeping social concerns in view' (12), 'market-friendly land use planning' (19) and 'buoyant, broad-based growth of employment, incomes and investment' (9).[36] The cornerstone of contemporary development discourse that development is only achievable on the basis of pro-poor economic growth therefore finds special resonance in urban sector strategies.[37] For example the World Bank assesses the prospects of Mercosur cities as follows (Mercosur is the 'southern Common Market' between Argentina, Brazil, Paraguay, Uruguay and Venezuela):

We know that the increasing interdependence of markets for all kinds of goods and services within the global economy, within countries, regions and localities, has resulted in new patterns of production, consumption, demand and supply . . . Over the last ten years, 50 percent of the jobs in Buenos Aires have changed from one sector to another. These changing patterns are reflected in new prices: for capital, labour, technology, skills, land, infrastructure . . . These prices in turn mean that the local demand for labor, for example, will depend no longer on local markets, but on a range of international markets . . . These short-term changes may occur in export markets and in the tourism industry . . . if fewer Brazilian tourists visit Mercosur cities, demand for facilities is likely to be reduced. One interesting impact will be the reduction in municipal tax revenue from sources tied to tourism, such as hotel receipts, airport taxes, and other charges. It will also have an impact on the demand for labor in the tourism industry.[38]

There are few cities in Latin America and the Caribbean, let alone anywhere else, that can meet the World Bank's benchmarks. Cities often lack a technocratic 'order', often suffer from a certain dirtiness and compromise health in certain ways. They are very often unsafe. Some would argue that it is indeed the lack of apparent order that is the attraction. Indeed some authors argue that such 'middle-city urban cultures' that give full expression to the informality, flow and tapestry of urban life are critical to the development process.[39] But the key question, especially given the power of global players such as the World Bank, is: what do cities do to meet these benchmarks? What does a city like San Salvador do to its poor and low-income communities to make it 'attractive' to tourists? For Theroux, if the likes of San Salvador and Guatemala City

> were hosed down, all the shacks cleared and the people rehoused in tidy bungalows, the buildings painted, the stray dogs collared and fed, the children shoed, the refuse picked up in the parks, the soldiers pensioned off . . . and all the political prisoners released, those cities would, I think, begin to look a little like San José,[40]

an exceptional city, he concludes.

It is not the importance of economic development to the life of cities that is questionable. Clearly it is. Cities are largely founded on cash-based economies, and urban residents therefore need work, both formal and informal. Rather it is the primacy, in some cases exclusively so, of economic growth in cities that has resulted in biased and uneven approaches to development. As Berwari and Mutter argue, the 'dimensions of development and poverty alleviation obviously include more than economics and economic growth'.[41] Moving from an economic-centred view of development means moving beyond income-centric views of poverty and prosperity and challenging the 'literature of development theories that had occupied centre stage of economic thinking during the cold war and that is irrelevant'.[42]

Recycling places: heritage and the urban poor

The iconic image and presence of spectacular colonial architecture in the cities of the Americas is undeniably a powerful draw card. As Tannerfeldt and Ljung suggest, tourism is growing quickly in historic cities, and 'spectacular sites are obviously magnets that attract capital and business. In many places cultural heritage is the most important single asset, although its potential may not have been realized.'[43] And, as the Inter-American Development Bank's Eduardo Rojas records, there are many cities in Latin America that 'are blessed with a rich legacy of buildings, public spaces, and urban structure'.[44] In combination, pre-Columbian, colonial (both Spanish and Portuguese) and post-colonial industrial (late nineteenth-century) architecture provide for the requisites of urban-based tourism: urban heritage.[45]

The notion of 'cultural heritage' has been traditionally linked with ancient monuments and archaeology, and in the context of the discussion above more readily associated with rural resources than the 'contemporary fabric of globalization and urbanization'.[46] But 'culture' is by no means the preserve of rural communities. As Charles Landry notes of cities, however, whilst tourism 'feeds off culture', most tourism 'focuses on a narrow conception of culture – museums, galleries, theatre and shopping'[47] – rather than the rich cultural distinctiveness of individual cities. Nevertheless, the historical fabric of many cities also represents a considerable 'financial heritage':

> The role of local culture, and cultural heritage, in the debate on cities is important and deserves consideration in the aid debate as well. If we understand past investment in infrastructure, museums, public space, and other facilities as part of a wider definition of urban cultural heritage, we need to reconsider how the *patrimonio* can be valued and utilized as an economic, cultural, and social resource as well. This is far beyond the common argument about tourism, but it involves a serious examination of the flow of benefits that urban areas can receive from earlier investments.[48]

This attitude towards heritage has changed dramatically over the past thirty years, a change that was especially accelerated by the introduction of the UNESCO World Heritage List in 1972. The World Heritage List includes 830 properties forming part of the cultural and natural heritage, which the World Heritage Committee considers as having outstanding universal value. These include 644 cultural, 162 natural and 24 mixed properties in 138 states, and an increasing proportion of sites listed which are city centres and historic towns (approximately a quarter of all listings). Listing remains important and a goal of many historic centres not least because cultural tourism is becoming an increasingly significant component of the global tourism industry and a package of the new tourism industry in particular.

The presence of these historic tourism resources is not, however, without its own problems and tensions. Whilst the stress of urban expansion and change, together with the overall 'poverty' of many city centres, undercuts the conservation of this richly textured urban landscape, tourism is seen as a potential (and sustainable) mechanism in support of rehabilitation and renewal. However there are protracted issues in the balance between 'conserving' and 'preserving', and the need for these areas to work within the existing grain of urban life and economic activity (often characterised by a thriving informal economy) rather than catalysing gentrification and effectively 'cleansing' such areas of the vulnerable and urban poor. With the potential prize of economic boosterism offered by urban-based tourism, there is a clear tension to directing tourism activities and revenue to pro-poor tourism strategies and initiatives that support the presence and livelihoods of the urban poor. Sylvio Mutal, an international consultant for the organisation World Historic Cities, addresses this relationship between urban heritage and urban tourism:

> in some areas of the world, some projects concerning historic city development may well have created new problems for local populations e.g. excessive stress on 'tourism'. Tourism and other income generating focus need to be put together. Tourism cannot be the magic key for investment. Our challenge is to reverse negative tendencies and create a symbiotic relationship between urban development and heritage preservation for *the improvement of social economic conditions of residents of all walks of life*. As one of the aspects of historic city life is urban poverty and social exclusion, there is a trend to improve the quality of life of inhabitants of cities through historic city programmes in ways that at the same time improve the urban environmental context and preserve and enhance cultural values, conserving adequately the built urban cultural heritage.[49] (Emphasis in original)

Squaring this complex circle is not easy. As the US commentator Mike Davis observes, 'Whatever their former splendour, most of Guatemala City's *palomares*, Rio's *avenidas*, Buenos Aires' and Santiago's *conventillos*, Quito's *quintas*, and Old Havana's *cuarterias* are now dangerously dilapidated and massively over-crowded'.[50] The arrival of comfortable, even bijou, cafés, bars and upmarket craft emporiums in cities as diverse as Quito, Antigua and Cartagena generally come at a price for the urban poor. And exactly the same tensions and contradictions are experienced by these new tourism urban resources as are experienced by new tourism rural resources in that 'tourism has a tendency to turn cities into museums, often compromising authenticity and either expelling the inhabitants or turning them into exotic exhibition objects for the tourists – as if they were in a zoo'.[51] Acknowledging the complex pressures of contemporary cities and the occupancy of the urban poor, however, the Director of the UNESCO World Heritage Centre, Francesco Bandarin, reportedly commented: '200 cities = 200 headaches'.[52]

If a kind of gentrification creep is detectable through alternative tourism in 'natural' ecological rich areas, it tends to be institutionalised and economically sanctioned in Latin American cities as elsewhere, requiring the full participation of property owners and the private sector. Like physically 'run-down' neighbourhoods anywhere, tarting-up comes at a high price both economically and socially. Even the Inter-American Development Bank, at the vanguard of neoliberal financed change in Latin America and the Caribbean, signals the downside. As the Urban Development Specialist Eduardo Rojas at the Bank comments: 'while gentrification benefits municipalities and landowners, it tends to expel low-income families and less profitable economic activities from the area. The poor lose access to cheap housing and to the economic and social opportunities offered by a downtown location.'[53]

There are, of course, examples of the successful marriage of heritage, tourism and pro-poor initiatives. As Gutiérrez observes,

> strategies for functional regeneration and recycling, the generation of diversifying activities, the encouragement of cultural tourism, and other means of improving the quality of life of the people have made outstanding contributions to the resurgence of historic centres, as can be seen in the case of Quito or more recently in Lima.[54]

In Havana too, the quality of life is being raised as restoration of its old centre is carried out. In this case – see Figure 6.1 – the local inhabitants have not been displaced by the renovation which represents something of an exception in the LAC countries. But in general in the LAC countries, the undertow of restoration and rehabilitation tends to be characterised by the same tendencies and tensions: spiralling land prices and land speculation, prices that are too high for the urban poor and clearances (or 'relocations' as they are known in the trade), often to those mile-upon-mile of unremitting misery that Theroux laments. As Hermer asserts, there are World Heritage listings that have 'saved the cultural heritage at the cost of aggravated social exclusion'.[55] But gentrification is regarded by most as an unavoidable by-product of urban conservation efforts,[56] and a resignation that 'comprehensive restoration might result in gentrification – that the rents go up so that only the well-off can afford to live there'.[57]

Pro-poor city tourism? The battle for land

If 'responsibility in tourism' were to accentuate the need for new tourism (and its continued growth in cities) to be what could crudely be labelled 'socially neutral' – that is not having adverse effects on the urban poor – then the more vexed question of whether it should positively promote, enhance and support the needs of the urban poor becomes, as Goran Tannerfeldt, a former senior urban development adviser to the Swedish development agency (Sida), provocatively asks:

> why should we preserve the cultural heritage and why should we support this in development co-operation? Will it contribute to the eradication of poverty? Will it improve gender conditions, promote human rights and democracy or any of the other objectives for development co-operation? My answer is: So what? Maybe it is something that is justified on its own merits and not in relation to other objectives.[58]

Figure 6.1 Tourist groups guided around the restoration of Old Havana
Photo: Martin Mowforth

In reality, there are few systematic studies on either the impact of tourism on Third World cities or the potential of tourism for reducing poverty through pro-poor planning and strategy development.[59] For example, and stepping outside the Latin America and Caribbean Region for one moment, in the absence of urban pro-poor tourism research, the major conclusion of a regional workshop bringing together mayors and officials from the Asia-Pacific Region 'was the need for a significant paradigm shift in the way tourism development occurs'.[60] This necessitates that 'tourism officials must shift from a situation where tourism arrivals are the primary indicator of tourism success to one concerned with a sustainable approach which improves conditions for the poor through tourism development'.[61] Nevertheless, the potential of cities for reducing poverty is increasingly acknowledged. The Asian Development Bank (ADB) for example reports: 'Properly planned and managed, urban tourism can be a significant tool for pro-poor urban development. It is labor-intensive and a recognized job creator. It does not require high academic skills. Vocational and basic skills, which the poor can easily acquire, are sufficient.'[62]

If pro-poor tourism were to offer anything, one could think of nothing better than the right to reside in the city and the right to an adequate home. For those inhabiting the expanding cities and towns of Latin America and the Caribbean (and other areas of the world), especially those crowded into the chaotic informal settlements on the outskirts and traditionally less desirable parts of the city, home is often far from safe, secure and settled. So-called shanties and slums are an anathema to tourism. Cities want to present the best face (as the discussion above suggests), to shift the crap that sullies the image. It is these pressures and perceived opportunities that work towards eviction and displacement. The vulnerable poor, despite the ingenuity of adapting to the harshness of cash-based urban economies, have few rights and little security:

> The idea of remoção (removal) is nothing new to Rio de Janeiro, where nearly 20% of the population, a million people, now live in about 750 slums. During the 19th century, town planners forced thousands from the slums of central Rio in a bid to turn the city into a 'tropical Paris' . . . Now in a campaign by sections of the Brazilian media, the question of favela removal has been thrust back on the political agenda . . . Sectors of the tourist industry also back removal. With Rio hosting the Pan American Games in 2007, several partial removals are being planned . . . 'I'm absolutely in favour of removal,' the president of the Brazilian Association of Travel Agents, Carlos Alberto Ferreira, recently told Globo newspaper, arguing that without the favelas blighting the landscape, tourism levels would rise, the profits of which could be channelled into fighting poverty.[63]

Whether alternative and more responsible approaches to (urban) tourism can do anything to redress this imbalance of power is a moot point. What it can do is to appreciate the cultural vibrancy and organisation of those areas that are a precarious *home* to the urban poor. The right to stay on land and to a home – rights that westerners have largely come to take for granted – is the subject of heated debate. It is a debate that has filtered into thinking and practice in responsible tourism largely as a product of the evidence of eviction and displacement of local communities in rural areas in response to the expansion of ecotourism projects. Equally, however, there is a vociferous neoliberal advocacy of the significance of property ownership to the development process. This is no more heated than in Peru, home of the international celebrity economist, Hernando de Soto.

De Soto's work is particularly influential with key multilateral development agencies, including the World Bank. It is perhaps a 'window' into some of the most fundamental development debates (and arguably could be read as a manifesto for the continued growth and expansion of tourism as one of the key sectors of potential Third World development). In accepting that capitalist economies provide the only demonstrable and serious option for development in the post-cold war era, Hernando de Soto argues in *The Mystery of Capital: Why Capitalism Triumphs in the West and Fails Everywhere Else*[64] that the Third World is held back by an inability to produce capital. In a nutshell de Soto argues that development is forestalled in the Third World because of the retardation of capitalism there, crucially 'not because international globalization is failing but because developing . . . nations have been unable to "globalize" capital within their own countries'.[65]

The centrepiece of de Soto's case rests on the production of capital and the fundamental significance of formal, legal, property systems. In countries without formal property titling where the ownership of assets (the most significant of which is physical buildings) is difficult if not impossible to trace and validate, de Soto argues that most resources are financially invisible and cannot therefore be converted into capital. As de Soto argues, it is 'formal property that provides the process, the forms and the rules that fix assets in a condition that allows us to realize them as active capital'.[66] In effect there are vast amounts of 'dead capital' which could otherwise be used to assist development; de Soto's research team estimate that the 'real estate held but not legally owned by the poor of the Third World and former communist nations is at least $9.3 trillion',[67] a figure that dramatically outstrips, for example, the total direct foreign investment in these countries between 1989 and 1999. By contrast, capitalism in the West has triumphed precisely because of the existence of 'one formal representational system'[68] that records and regulates most assets.

Most stark is de Soto's unswerving belief that, if capitalism has the chance to spread and develop in the Third World, it will deliver development. Although de Soto acknowledges the inequitable nature of contemporary capitalism in developing countries, and indeed denies what might appear as the credentials of a 'diehard capitalist', it is in his final analysis 'the only game in town'.[69] If governments are willing to accept the 'property manifesto', the critical problems of economic and social apartheid are 'relatively easy to correct'.[70] In an approach reminiscent of the pro-poor development paradigm, he confidently predicts from the 'perspective of the poor' that 'Everyone will benefit from globalizing capitalism within a country, but the most obvious and largest beneficiary will be the poor'.[71] We would suggest, on the other hand, that if such an approach were taken up with gusto by national and city authorities eager to transform their cities into modern archetypes and flushed with tourism and tourism facilities, the prospect for the poor would look precarious at best, bleak at worst.

The seductiveness of the approach taken by the likes of de Soto lies in part in their conformity to economic tradition (that is in their respect for economic growth and capitalist organisation) and in part in their simplicity, universality and supposed promotion of the economic 'missing link'. They are to development economics what travel books and self-help books are to western consumers. There is a belief that incentives are capable of overcoming the barriers to development. For de Soto this means that formal property offers the 'means to motivate people to create real additional usable value'.[72] For de Soto, the implication is that capitalism has produced equitable and healthy societies in the West. One thing is clear, there is a widely held belief in economic growth as fundamental to the wellbeing of the poor and the progress of development, and little questioning of the need to slow down and redirect growth, with an attendant, scaling down of our comparatively luxurious, free and consumptionist western lifestyles.

Trivialising the poor or taking control? *Favela* tours

'Sensitising' tours already exist in some cities focusing on the exposure of the 'other' side of cities, or perhaps cashing in on the notoriety of some city areas. The cultural difference of such areas is the potential source of attraction, and tourism seeks to exploit this difference. The activity is often justified by the trickle-down assumption of economic growth and benefit. Much like rural-based tourism, however, the fundamental questions remain about access to, ownership and control of resources and the degree of power vested in poorer communities as determinants of pro-poor development potential. There is scant literature which addresses and details the dynamics of new and supposedly responsible forms of tourism in urban areas, and the problems and prospects that arise from them, especially in the LAC countries. This is one of the reasons why we focus in this section on the innovative work of Deborah Dwek in Rio de Janeiro.

In Brazil, slum areas of cities are termed *favelas*. Many *favelas* have their origins in overnight migrations of large numbers of people from rural areas to the cities, having been dispossessed of their land or made jobless. In today's terminology, their original inhabitants would be referred to as economic migrants. Initially, their houses fit the descriptions given by Theroux earlier in this chapter, being flimsy encampments using materials of plastic, tin sheeting and even cardboard.

In some settings *favelas* would be known as slums, shanties, *campamentos*, squatter camps. At first, they are set up without any services, but if they escape clearance by the authorities and as they become established, house improvements take place, corner shops may be set up, electricity lines hijacked and later even workshops and offices established; so that eventually enough pressure is exerted on or is felt by the authorities for services to be extended into the *favela*. As Deborah Dwek's work on *favela* tourism makes clear, most of today's *favelas* do not deserve the label of 'slum', and many *favelados*[73] would object to its use to describe their residential area.[74]

Despite this, it is also clear that these are areas which in popular perception have become associated with poverty, deprivation, crime and violence, and the extraordinary rise to prominence of cocaine and its introduction into the *favelas* at the end of the 1970s and early 1980s led to the establishment of drug gangs and organised crime networks which still dominate the image of the *favelas* held by outsiders. This image leads to a mixture of prejudice and fear about the *favelas* and *favelados*, which is felt at least as much by Brazilians as it is by foreigners. As Dwek remarks:

> I found that this attitude which revealed both prejudice and fear was representative of the attitude of many middle-class Brazilians who, while cohabiting and interacting with *favelados* on a daily basis, confined this contact to a hierarchical relationship with maids, porters or others within the service industries. Few had been inside a *favela* . . ., associating them immediately with drugs and danger. Most thought it was still too dangerous to go into Rocinha, a tribute in part to the powerful job the media has done in promoting this view.[75]

It is paradoxical that the *favelas* are also so strongly associated with a number of positive aspects of Brazil's image, especially the samba and carnivals; and it is this mixture of images – fear and danger, yet excitement and sensation – which makes them an attraction to visitors. 'While Brazil uses popular culture to successfully market her image on the world stage, prejudice against *favelados* within Brazil runs deep.'[76] Most visitors actually aim to experience a sense of threat and danger, whether real or imagined.

The *favela* tour is a recent feature of the Brazilian tourism industry, set up to exploit this 'sexy' image, as Dwek calls it, and what has become known as '*favela chic*'. Dwek's research into *favela* tourism was carried out in 2004 in *favelas* in Rio de Janeiro. Such research is relatively new; indeed, such experiences are relatively new too, although officially the *favela* of Rocinha in Rio de Janeiro has been receiving tourists since 1992. It is commonly perceived that *favela* tours are run by residents of the *favela* and are thus an indication of local entrepreneurial spirit rising out of the hardship of life. As we pointed out in Chapter 3, however, Deborah Dwek's work has shown that most of the tours are actually managed by outsiders and surprisingly few residents even act as guides. Despite this, it is one of Dwek's conclusions that most residents viewed the tours in an extremely positive light, and there is little doubt that some residents gain financially from the interchange. In fact, Dwek reports that a number of artists have developed a degree of dependence on the tours for selling their work and that their economic livelihood becomes precarious if the tours are cancelled. Despite the trickle-down of economic benefits to a few individuals, Dwek found that there were no returns to the community as a whole.

Box 6.1 presents a selection of observations on *favela* tourism made by Dwek. Her report is an important and insightful one, pointing out many of the contradictions of supposedly responsible forms of tourism in a setting which can be expected to highlight such contradictions. She suggests that 'there is still a stratum of tourists who find the idea of these tours ethically dubious',[77] a reference to the voyeuristic nature of such tours which is hinted at in several of the observations made in Box 6.1. The ethical doubts probably also reflect the concern of some critics regarding the search for the 'authentic other' by the modern, or perhaps it should be postmodern, tourist. The *favela* tour is the chance for many of the middle-class tourists and backpackers (identified by Dwek as the major groups of visitors using

---

Box 6.1   Dwek on favela tours

This external fascination of the West with the favela has always been about projecting outsiders' perceptions and values on to this world. (p. 11)

I was witness to this safari-style viewing on my *Jeep Tour* with an extended family of American tourists complete with cameras. After entering a house of one of the local families in Parque da Cidade favela a 'typical scene' was pointed out to us of a family relaxing on their balcony below and I felt our role as 'gazer' was made uncomfortably clear as we were given the impression of viewing the natives in their natural habitat, almost as if we should not be making too much noise in case we disturbed them! (p. 21)

If the police ever stop him while he is guiding a tour it only adds to the excitement for the tourists . . . They also clearly liked the 'danger' element – which Fantozzi played up to for the thrill-seeking tourist – emphasising the 'underworld' elements of the favela such as drug-related graffiti and the traffickers themselves – some of whom the tourists were convinced were following them. (p. 21)

One particular incident on Fantozzi's tour struck me as being nothing less than voyeuristic as well as crass and exploitative. An old lady from Rocinha was asked to pick one of the male tourists to kiss, and as she played up to her audience everyone laughed which reminded me of a circus act; stripping the lady of any dignity and only serving to reinforce the stereotype of the 'noble savage'. (p. 22)

This superior and patronising attitude that Ludemir refers to, was displayed by some of the tourists who were frequently surprised at how friendly, well-behaved and receptive the *favelados* were, having expected them to be more hostile as well as noting how they seemed to treat each other very well, even sharing their food among themselves. It was as if they were another species of people who belonged to a world unlike anything they knew. (pp. 25–6)

Source: Deborah Dwek (2004) 'Favela Tourism: Innocent Fascination or Inevitable Exploitation?', MA dissertation, Institute of Latin American Studies, University of London.

---

the tours which are on offer to the *favelas*) to get closer to the 'real' Rio. 'Today's tourist, coming predominantly from the post-industrialised and post-modern western world, travels in search of the antithesis to capitalism and modernity',[78] and it is this search that leads them into a *favela* tour. She found that a number of tour companies made this search for the 'other' and for the 'real' Rio a major focus of their tours, whilst others gave a more objective portrayal of the areas. Of course, the very existence of tourism in the *favelas* alters the nature of the interchange and experience for both the visitor and the visited, bringing into question the validity of the authenticity of the experience.

Despite this, Dwek concludes that such tours can be an effective way of dispelling prejudices about the *favelas*. Her major conclusions are summarised as follows:

- The majority of *favela* residents have a positive view of the *favela* tours
- *Favela* residents have their own ideas about how to exploit tourism, but 'do not appear to have a voice with which to express them or any power to effect change'[79]
- Initiatives which spring from the *favelados* themselves will produce the greatest economic returns for the communities
- Realistically, tourism will be an option only for a few *favelas* in Rio, and even in those cases it is unlikely that it will serve as a model for economic development
- Investment and development in the *favelas* should come from the government rather than from tourists
- Despite the problems of tourism's association with voyeurism and imperialism, it appears to be an effective mechanism for breaking down social barriers and stigmas
- Perhaps the most significant results that 'responsible tours to *favelas* can generate is a politicisation and awareness for tourists of the social problems that Brazil is facing';[80] and
- The regulation of tours by external bodies will be one way of ensuring that the *favelas* are not misrepresented or exploited.

Most of these conclusions are either supported or alluded to by Tom Phillips in his report on the Brazilian government's 2007 plan to build tourist guesthouses (*pousadas*) in Rocinha, Rio de Janeiro's largest shantytown.[81] Box 6.2 includes selected extracts from Phillips's article to give the reader a flavour of the Brazilian government scheme. Certainly it can be seen that the plan has generated hope and enthusiasm amongst the residents of the area. The importance of integrating tourism initiatives within the general thrust of development is also a point well made. But, as the last comment in Box 6.2 makes clear, *favela* tourism is unlikely on its own to be the cure for poverty, crime and social problems.

Concluding remarks

Although cities have been traditionally associated with tourism, there is relatively little written on urban tourism and the impacts of tourism on cities in the so-called 'developing world'. Whilst the environmental and cultural impacts of tourism in rural, coastal and natural areas are well documented and critiqued, city-based tourism receives less attention and for the most part the coverage is limited to the size, growth and economic impact and potential of tourism. Equally, whilst alternative community-friendly travel guides are on the increase, there is little

---

Box 6.2   Tourist guesthouses in Rio's shantytowns

The Brazilian government has announced multimillion pound plans to build tourist guesthouses – *pousadas* – in one of the most notoriously violent corners of Rio de Janeiro.

President Luiz Inacio Lula da Silva revealed the plans on Thursday as part of a development project which also includes the construction of roads, creches, hospitals and a convention centre in Rocinha, Rio's largest shantytown . . .

The *pousadas* . . . are expected to be located in Laboriaux, one of the highest sections of the *favela*. The area boasts spectacular views over Rio's undulating landscape but is also known for shoot-outs between drug traffickers and police and is located near clandestine cemeteries used by traffickers to dispose of their enemies.

Yesterday the mood on the Ladeira do Laboriaux – a steep incline that leads into the shantytown – was buoyant. 'Gringos?' said Cristiane Felix de Lima, sitting in her husband's bar, Seven Lives, in the hilltop *favela*. 'It will be good for business.'

Ricardo Gouveia, an architect and human rights activist, said the new attempts to combine 'land rights, urban redevelopment and social projects' represented a significant step towards improving living conditions in Rio's biggest shantytown, home to over 100,000 impoverished Brazilians.

Few deny that developing the sprawling shantytown represents a huge challenge. 'To sort out Laboriaux and Rocinha, you have to look at all the problems,' said community leader Paulo Sergio Gomes as he stood in a shack constructed out of branches and abandoned wardrobe doors. Beside him three naked children – aged five, six and seven – played in the dust. Their mother, who works 15 hour days, had abandoned them at home . . . 'Most of the houses around here are even worse than that one,' Mr Gomes added, as a teenager hurtled past on a motorbike, a pistol tucked in his belt. 'We need more than half a dozen *pousadas* to fix things.'

Source: Extracts from Tom Phillips (2007) 'Brazil to Build Tourist Guesthouses in the Heart of Rio's Shantytowns', London: *The Guardian*, 20 January.

---

coverage of these topics, in spite of the fact that cities house the majority of the region's population. This is surprising in that a crop of mainstream and alternative travel guides have emerged focusing on key cities (including the *Cities of the Imagination* guides). This is especially notable in Latin America with a handful of outstanding cities, whose growth and global recognition have quintessentially been related to tourism (from Rio to Acapulco) and are heavily implicated in the political economy of the region (such as Havana's former playground image).

It is logical perhaps that recent attempts to use new and responsible forms of tourism as 'developmental' tools or activities have found favour in community-based rural tourism and that pro-poor developments have followed a trend of focusing almost exclusively on rural-based tourism.[82] Where there is attention to cities, it tends to be in relation to those aspects or products of tourism that ignore urban problems and the potential role of responsible tourism. Admittedly, the discussion of low-income cities and 'pro-poor tourism' is a recent one, but, given the increasing urbanisation of poverty[83] and the inherent attractiveness of cities, it remains surprising that there is this relative silence on the pro-poor opportunities in urban-based tourism activities.

Notes

1 Paul Theroux (1979) *The Old Patagonian Express: By Train Through the Americas*, London: Penguin Books, p. 129.
2 Babar Mumtaz (2001) 'Slums Are Good for You', *Habitat Debate*, 17, 3, September.
3 Colantonio, A. and Potter, R. (2006) 'The Rise of Urban Tourism in Havana since 1989' *Geography* 91, 1: 23–33: 23.
4 UN-HABITAT (2003) *The Challenge of Slums: Global Report on Human Settlements 2003*, London: Earthscan.
5 Angotti, T. (1995) 'The Latin American Metropolis and the Growth of Inequality', *NACLA Report on the Americas*, XXVIII, 4: 13–18.
6 *Ibid.* p. 13.
7 Paul Theroux (1979), p. 161.
8 Miles Warde (1992) 'Where Thieves Often Prosper', London: *The Guardian Weekend*, 14 November: 51–2.
9 Isobel Wolff (1991) 'Weekending in El Salvador', London: *Independent on Sunday*, 27 October: 59, 61.
10 World Bank (2003) *A Resource Guide for Municipalities: Community-Based Crime and Violence Prevention in Urban Latin America*, Washington DC: World Bank.
11 *Ibid.*, p. 7.
12 Guerrero, R. (1999) *Violence Prevention – Technical Note 8: Violence Control at the Municipal Level*, Sustainable Development Department, Washington DC: Inter-American Development Bank.
13 Winton, A. (2004) 'Urban Violence: A Guide to the Literature', *Environment & Urbanization* 16, 2: 165–84.
14 *Ibid.*, p. 167.
15 *Ibid.*, p. 168.
16 Winton, A. (2004).
17 Caroline Moser (2004) 'Urban Violence and Insecurity: An Introductory Roadmap', *Environment & Urbanization* 16, 2: 3–16.
18 Maggie Black (1994) *Mega-slums: The Coming Sanitary Crisis*, London: WaterAid, March.
19 As Atkinson suggests in the context of global urbanisation, it is increasingly perverse that 'development urbanists' should continue finding themselves working at the 'margins of the broad field of development' (2002: 178) and that until recently there has been an 'attrition' of urban specialists from key institutions (World Bank 2000: 16). Only recently has the historically weak United Nations Centre for Human Settlements been transformed into a full UN Programme (to United Nations Human Settlements Programme, known as 'UN-Habitat') in spite of the fractional UN and other lending to human settlements expenditure noted in the United Nations Conference on

Environment and Development Agenda 21 (UNCED 1992), and many bilateral aid programmes have ignored or marginalised urban issues within broader non-sectoral programmes. 'Urban aid' fails even to get a mention in the popular annual Reality of Aid series (Satterthwaite 2001; Cohen 2001). Discussions with Overseas Regional Programme Managers in the UK volunteer sending agency Voluntary Service Overseas also listed the proximity of volunteers to country programme offices and the possibility of volunteers 'bothering field staff', located mainly in the capital cities, as a reason for avoiding work in urban areas (Personal Notes, 1995). References: Atkinson, A. (2002) 'International Co-operation in Pursuit of Sustainable Cities', in Westendorff, D. and Eade, D. (eds) *Development and Cities*, Oxford: Oxfam, p. 177–203; World Bank (2000) *Urban and Local Government Strategy*, Washington DC: World Bank; United Nations Conference on Environment and Development Agenda 21 (UNCED, 1992); Satterthwaite, D. (2001) 'Reducing Urban Poverty: Constraints on the Effectiveness of Aid Agencies and Development Banks and Some Suggestions for Change', *Environment & Urbanization* 13, 1: 137–57; Cohen, M. (2001) 'Urban Assistance and the Material World: Learning by Doing at the World Bank', *Environment & Urbanization* 13, 1: 37–60.

20 Gareth McLean (2006) 'Where We're Headed' *The Guardian Weekend*, 1 April: 38–43.

21 Angotti, T. (1995), p. 16.

22 Hague, C., Wakely, P., Crespin, J. and Jasko, C. (2006) *Making Planning Work: A Guide to Approaches and Skills*, Rugby: Intermediate Technology Publications Ltd: p. 86.

23 Deborah Eade (2002) 'Preface' in Eade, D. and Westendorff, D. (eds) *Development and Cities*, Oxford: Oxfam GB, p. xi.

24 Nederveen Pieters, J. (2001) *Development Theory: Deconstructions/Reconstructions*, London: Sage.

25 The ruralisation of development took its most dramatic manifestations in a continuum of practice that ranged from Tanzania's Ujamaa experiment to Pol Pot's urban clearances, both in part fuelled by a fear of urban parasitism and exploitation of rural peasantry. Schumacher suggested that the upper population of a city should be no more than half a million.

26 Ernst Schumacher (1974) *Small Is Beautiful: A Study of Economics as if People Mattered*, London: Abacus, p. 64.

27 This section draws on Joel Simon's account in *Endangered Mexico: An Environment on the Edge*, London, Latin America Bureau (1997).

28 Joel Simon (1997) *Endangered Mexico: An Environment on the Edge*, London: Latin America Bureau, p. 195.

29 Esteva, G. (1992) 'Development', in Wolfgang Sachs (ed.) *The Development Dictionary: A Guide to Knowledge as Power*. London: Zed Books.

30 Burgess, R., Carmona, M. and Kolstee, T. (1997) 'Contemporary Policies for Enablement and Participation: A Critical Review', in Burgess, R., Carmona, M. and Kolstee, T. (eds) *The Challenge of Sustainable Cities: Neoliberalism and Urban Strategies in Developing Countries*, London: Zed Books, p. 118.

31 See for example, Alfonso Salazar (1990) *Born to Die in Medellin*, CINEP, Colombia.

32 Sachs, J. (2005) *The End of Poverty: How We Can Make It Happen In Our Lifetime*, London: Penguin, p. 36.

33 World Bank (2000) *Urban and Local Government Strategy*, Washington DC: World Bank, p. 8.

34 *Ibid.*, p. 15.

35 Mila Freire and Richard Stren (eds) (2001) *The Challenge of Urban Government: Policies and Practices*, Washington DC: The International Bank for Reconstruction and Development/The World Bank, p. 13.

36 World Bank (2000) *Urban and Local Government Strategy*, Washington DC: World Bank.

37 As Mitlin (Mitlin, D. (2002) 'Sustaining Markets or Sustaining Poverty Reduction?' *Environment & Urbanisation* 14, 1: 173–77) forcefully argues this 'fetishism of the market' is also evident in the tendency to graft financial sustainability (variously referred to as bankability, full cost recovery, privatisation, private sector involvement and so on) as one of the principal measures of development success (Mitlin, 2002), and that the most enduring characteristics of globalisation of the extension of markets may in retrospect be seen as a 'somewhat incredulous faith in markets' (2002: 174).

38 Freire and Stren (2001).

39 Samuels, J. (ed.) (2005) *Removing Unfreedoms: Citizens as Agents of Change in Urban Development*, Rugby: ITDG Publishing.

40 Paul Theroux (1979).

41 Berwari, N. and Mutter, M. (2005) 'Introduction', in Samuels, J. (ed.) *Removing Unfreedoms: Citizens as Agents of Change in Urban Development*, Rugby: ITDG Publishing.

42 Samuels, J. (2005).

43 Tannerfeldt, G. and Ljung, P. (2006) *More Urban Less Poor: An Introduction to Urban Development and Management*, London: Earthscan, p. 106.

44 Rojas, E. (2002) *Urban Heritage Conservation in Latin America and the Caribbean: A Task for All Actors*, Washington DC: Inter-American Development Bank.

45 In Latin America and the Caribbean this includes numerous sites and urban areas variously described as 'historic centres' (unmistakably characterised by the presence of a Plaza Mayor that from the sixteenth century became the centrepiece of Spanish colonial towns and which are 'without doubt favoured heritage sites' (Gutiérrez 2001)) and 'historic towns' probably representing 'one of the most important heritage resources on the American continent' (Gutiérrez 2001). A final category of 'integrated architectural groups', foremost of which is the fortifications of the Caribbean region, are already on the UNESCO World Heritage list (including Havana, Cartagena de Indias, San Juan de Puerto Rico, and Portobelo and, Gutiérrez suggests, is closely tied in with 'cultural tourism circuits' (2001). Gutiérrez, R. (2001) 'The Urban Architectural Heritage of Latin America', available from http://www.icomos.org/studies/latin-towns.htm, accessed July 2006.

46 Hermer, O. (2005) 'Aches to Assets' in Sida (ed.) *Urban Assets: Cultural Heritage as a Tool for Development*, Stockholm: Sida, pp. 5–10.

47 Landry, C. (2000) *The Creative City: A Toolkit for Urban Innovators*, London: Comedia and Earthscan, p. 9.

48 Cohen, M. (2004) 'Reframing Urban Assistance: Scale, Ambition and Possibility', Urban Update No. 5, Woodrow Wilson International Centre for Scholars, Washington DC, p. 8.

49 Sylvio Mutal (2005) 'Heritage and Urban Development – Considerations on New Approaches to Heritage', in Sida: *Urban Assets: Cultural Heritage as a Tool for Development*, Stockholm: Sida, pp. 23–9.

50 Mike Davis (2006) *Planet of Slums*, London: Verso, p. 32.

51 Hermer, O. (2005).

52 *Ibid*.

53 Eduardo Rojas (1999) *Old Cities, New Assets: Preserving Latin America's Urban Heritage*, Inter-American Development Bank, p. 17.

54 Gutiérrez, R. (2001).

55 Hermer, O. (2005), p. 6.

56 Sida (2005) *Urban Assets: Cultural Heritage as a Tool for Development*, Stockholm: Sida.

57 Tannerfeldt and Ljung (2006), p. 108.

58 Goran Tannerfeldt (2004) *A Future for the Past: Historic Cities in Development*, Stockholm: Sida, pp. 12–13.
59 This is therefore a key area for future innovative research. Where such studies do exist, they tend to conflate correlation and causality – for example, Colantonio and Potter (2006) – and are inconclusive.
60 Jamieson, W. (2002) *Workshop Report: Regional Workshop – Urban Tourism and Poverty Reduction*, Colombo, Sri Lanka, November 20–2, ESCAP/CITYNET.
61 *Ibid.*, p. 7.
62 Asian Development Bank (2006) 'Urban Poverty. Tourism: More than Sight-seeing', http://www.adb.org (accessed July 2006).
63 Phillips, T. (2005) 'Blood, Sweat and Fears in Favelas of Rio', London: *The Guardian*, 29 October, p. 11.
64 Hernando de Soto (2000) *On the Mystery of Capital: Why Capitalism Triumphs in the West and Fails Everywhere Else*, Black Swan: London.
65 *Ibid.*, p. 219.
66 *Ibid.*, p. 44.
67 *Ibid.*, p. 32.
68 *Ibid.*, p. 50.
69 *Ibid.*, p. 241.
70 *Ibid.*, p. 241.
71 *Ibid.*, p. 201.
72 *Ibid.*, p. 228.
73 Favelado: inhabitant of a favela.
74 Deborah Dwek (2004) 'Favela Tourism: Innocent Fascination or Inevitable Exploitation?', MA dissertation, London: Institute of Latin American Studies.
75 *Ibid.*, p. 24.
76 *Ibid.*, p. 24.
77 *Ibid.*, p. 14.
78 *Ibid.*, p. 22.
79 *Ibid.*, p. 22.
80 *Ibid.*, p. 31.
81 Tom Phillips (2007) 'Brazil to Build Tourist Guesthouses in the Heart of Rio's Shantytowns', London: *The Guardian*, 20 January.
82 For example, 'product development plans that include rural/cultural/adventure/ community tourism, or other products suitable to development in poorer areas and by small-scale entrepreneurs and for which there is a defined market', Pro Poor Tourism Partnership (2004) Sheet No. 8, Policy instruments supporting PPT.
83 United Nations Centre for Human Settlements (UNCHS – Habitat) (1996) *An Urbanizing World: Global Report on Human Settlements 1996*, Oxford: Oxford University Press. UNCHS – Habitat (2001) *Cities in a Globalizing World: Global Report on Human Settlements 2001*, London: Earthscan.

7 Sexual exploitation through tourism

The relationship between sex and tourism has always been just below the surface, represented in the early days of the tourism industry by the smutty seaside postcards, and almost universally in recent decades by the advertisements that feature scantily clad women and men on a beach or sipping martinis by a pool. Tourism is an industry whose central themes are enjoyment and relaxation, themes which are also, at least from some standpoints, associated with sex. It is important therefore to distinguish between sex tourism and sex within tourism. The former embodies the exploitation of one group by another and the exercise of power by some over others. The latter reminds us that in itself sex is not necessarily disagreeable and is as likely to be acted upon consensually as through manipulation or coercion. We make it clear at the outset of this chapter, then, that our coverage here is of the former, the horror of whose effects is illustrated by the following description.

> In the area around El Belén in Tegucigalpa, the cheap, run-down bars offer pre-pubescent girls for unprotected sex for less than three dollars. No condoms, no safe sex. With 40,000 estimated cases, Honduras has the highest HIV rate in all of Central America. Most of these people are developing full-blown AIDS between the ages of 21 and 29, which means that most are being infected between the ages of 12 and 18.[1]

Throughout this book we have illustrated the belief that tourism acts as a channel through which power relationships are played out: the tourist placing his or her order with the waiter or waitress at its simplest; the First World tour operator dictating conditions to the local operator in Latin America; the cruise ship liner demanding instant facilities from the Caribbean island port authority; the international environmental organisation dictating to the locals how services should be provided; the porter carrying the trekker's 'necessities'; and so on. Sex can also be a channel through which power – usually but not always the physical and/or economic power of men over women – is transmitted. And tourism offers to men the possibility of living out sexual fantasies with the advantage of anonymity, an advantage not normally available in the home environment. Thus, in the same way that the young football fan can become an anonymous lager lout when away from home, so the fantasist can become an anonymous sexual predator when touring abroad. In such circumstances, sex represents an act of power and dominance for one and subservience for another.

But the issues associated with sex tourism are more complex than the simple relationship of physical dominance between men and women would suggest. The power differential is made all the more significant by the fact that here we are dealing with male visitors from the rich world, the First World, exploiting females from the destination country, in Latin America and the Caribbean, the Third World. The relationship is all the more unequal for the fact that in many cases it is children who are the exploited. But it is important that we set the incidence of sex tourism in the context of its scale. What proportion of international tourists does it involve? Are the majority of its users local or foreign? Is it organised or informal? The issues are complicated further by the incidence of women visitors seeking sex from local men. How is power transmitted in such a case? Through sex or through wealth? Then the emergence of the lesbian, gay, bisexual and transgender (LGBT – this is the way many sexual minority organisations choose to describe themselves) community adds a new dimension to the study of the issues of sex and tourism, and highlights the importance of gender and sexual identity in defining the relationships of power between individuals within society.

We look first at the bases for the prevailing expectations and attitudes towards sex and gender roles in society and how these can affect tourism. This is followed by an investigation into the phenomenon of sex tourism, a tourist activity on the increase in the LAC countries. Finally, in a book dealing with issues of responsibility, it is crucial that we examine the health and education issues associated with sex tourism.

Machismo, gender and sexual identity in Latin America and the Caribbean

The LAC countries are hardly the only ones where gender inequalities are rife and where attitudes of machismo are a prevailing determinant of the social and economic relationships between the sexes. It is an enduring feature of such relationships throughout the world. But it is a subcontinent often characterised by its machismo. So, what is machismo? How does it affect social and economic relations between the sexes? And how are these effects translated into behaviour patterns in the region? Moreover, are these effects and patterns reflected in the field of tourism?

In the words of Patrick Welsh, machismo is in effect:

> an ideology built upon the erroneous supposition that men, as a result of a natural phenomenon, are physically, intellectually and sexually superior to women, a concept that is instilled into both women and men from an early age. Consequently, men enjoy rights and privileges and have access to opportunities in society that are denied to women and a system of gender inequity is promoted and perpetuated.[2]

Welsh adds that it would be misleading and simplistic to believe that machismo in the LAC countries began with the arrival of the Spanish conquistadors more than 500 years ago. The promotion and maintenance of the idea of male domination in society is a universal phenomenon which crosses historic and cultural boundaries. In many indigenous societies women were considered to be the property of men. The currently prevailing system of gender relations in the Latin countries is a mixture of those inherited from the Hispanic and indigenous cultures, and in a majority of Caribbean societies those inherited from a mixture of the African, indigenous and European cultures, in that it stems from 'the birth of a new ethnic group through the rape of indigenous women, the derogation of autochthonous languages and the imposition of another language, religion and new social order'.[3]

Added to this mix were the misogynist attitudes of the Catholic Church through its pressure on women to emulate the Virgin Mary as a model of feminine integrity, requiring of all women purity and virginity. From its patriarchal viewpoint, the Catholic Church attempted to control the sexuality of women and can be seen as one of the major planks of machismo in Latin America.[4] Education systems throughout the subcontinent also have a tendency to reinforce attitudes and beliefs in the moral and social roles that should be played by women and men in society. Additionally, the family has been a key factor in this reinforcement of attitudes and roles, women being at least partly responsible for the inculcation of these values and attitudes within their children.

Whatever its origins, machismo has produced, or reproduced, the behaviour patterns of the conquistador, arbitrary and indifferent to public sentiment, scornful of women and which make the variety of sexual partners the measure of masculinity. It has assigned to women, as their 'natural' space, care of the house and the rearing of children. Men, on the other hand, are free of these responsibilities and therefore have a greater freedom of movement.

Over time within the family this translated into women bearing the responsibility for child care and housework and men bearing the responsibility for paid work. This division of responsibilities traditionally tied women to the home location and inhibited their access to opportunities for education and for paid work. Men, on the other hand, gained the kudos associated with paid work and the power associated with money.

Until recently these divisions and differences led to an enduring machismo throughout the LAC countries which has created the inequitable situation between the sexes within society that is evident today. It is a situation which reflects the traditional power structures built up over the centuries. The changes which have occurred over the last two decades of the twentieth century, whilst not necessarily bringing about a change in these traditional power structures, have brought into question the relationships of men to society and of women to society. They are changes prompted largely by the forces of economic globalisation. In many rural areas of the LAC countries the liberalisation of trade has undermined prices for primary products, has eroded the market for locally produced goods flooding it with imported goods and has severely reduced the viability of both subsistence farming and small-scale cash cropping. One of the principal results of these trends has been an increasing out-migration from rural areas to the cities. To many of these migrants, their predicament reflects a sense of failure. And in the cities the only new employment opportunities on offer from the same neoliberal economic policies are in the assembly and textile factories (*maquilas*) in free trade zones, work that is traditionally done predominantly by women.

Increasingly, then, men are now faced with the prospect of unemployment, no wage, yet further failure in their traditional role of provider for the family and a loss of a sense of importance and usefulness. Poverty and unemployment breed disillusionment and alienation. For many men, the solution to this situation is found by further migration (by legal or illegal means), this time to the wealthy neighbour to the north, the USA. For others the solution is found in 'typically male methods of dealing with crisis: increased alcohol consumption and violence'.[5] Except in a few cases, these changes do not appear to have reduced the effect of machismo; in fact, as the threat to the idealised family and its lifestyle norms increases, the standard macho attitudes and explanations for the situation may even strengthen.

The same forces have begun to alter the traditional role of women as the carer of the house and the children. The earlier roles remain, but increasingly women are also having to shoulder the burden of financial provider for the family. For many the only appropriate opportunities are provided by the *maquilas*, which pay marginally above slave wage rates and where labour conditions are the modern-day equivalent of the Victorian sweatshop factory. It is questionable whether this represents a liberation from *machismo* for women in the LAC countries; it is more likely for most that it represents an extra form of domination to add to those that women already suffer. But for men it represents a loss of influence, importance and, usefulness, and as Vanessa Baird explains, 'Autonomy – especially for women – is a threat to authoritarian and patriarchal control'.[6] Whether the extra burden of raising finance for the family really reflects an extra level of autonomy is dubious; but it is perceived that way by many men. Moreover, it is perceived by the political right and the fundamentalists with whom they are allied as a challenge to their value systems based on family and religion.[7]

Homosexuality also challenges the supposed 'naturalness' of traditional male and female roles in society and adds further to the threats facing unemployed and seemingly 'useless' men.[8] The standard homophobic objection to sexual behaviour that differs from heterosexual norms may be associated with the threat to the idealised family and its lifestyle norms. But it is also associated with gender which describes position with respect to society rather than 'sex' which is an attribute of your body and biology. As Vanessa Baird states:

> Deviation from heterosexual norms is threatening because it seems to challenge the conventional rules governing a person's sex, their sexual preferences and the general female and male roles in society. The assertion of homosexual identity clearly challenges the apparent naturalness of gender roles.[9]

The development of homophobic attitudes in men ties *machismo* even more strongly to the intolerance of the political right, although it is important to remember that *machismo* is not solely associated with one philosophical perspective or one political credo. In the LAC countries *machismo* is as likely to be encountered amongst the political left as it is amongst the right, although it is philosophically tied to the latter whilst it generally does not form part of the philosophy of the left.

Whichever direction it comes from, however, it can sour relations in and the development of the tourism industry. An example documented by Polly Pattullo illustrates a contradiction in the tourist industry's attitudes to gay and lesbian visitors. The case concerns the Cayman Islands' 1998 ban of a visit by a cruise ship chartered by gay organisations.

> The charter company then declared its intention to visit the Bahamas. The outcry in the Bahamas was sufficient to merit an intervention by Prime Minister Hubert Ingraham. In a powerful speech on national television, Ingraham was highly critical of the anti-gay lobby. He described the media coverage as 'a sea of bitter, poorly reasoned diatribe' and 'un-Christian'. While saying that the constitution guaranteed freedom of speech, he also said: 'Quite simply, it is not the role of the government to investigate and pass judgement on the sexual behaviour of consenting adults so long as their activity is conducted in private.' It was a brave and radical speech for any Caribbean politician to make, for the Caribbean's adherence to a fundamental, wrathful Christianity remains widespread, and homophobia is commonplace. Ingraham was also being pragmatic for in his speech he emphasised the importance of tourism – 'our economic lifeblood'; the Bahamas could ill-afford to exclude any visitor, whatever their sexuality.[10]

And so we arrive at the question of how *machismo* and the established gender roles affect the practice of tourism. The established gender roles determine, to a large extent, people's expectations, and expectations determine, to a large extent, the kind of job (and therefore role) that people seek. Managers and management boards of hotels appointing applicants to positions also look for people of the 'appropriate' gender, which is, to a large extent, predetermined. So cleaners and housemaids will be female – not exlusively, but largely – and drivers and porters will be mostly male. Various positions may be filled with people of either sex, such as receptionist, and waiters or waitresses. Such probabilities are obvious to us all and clearly signal that breaking out of traditionally expected roles is not an easy matter. But examples of iconoclasm do occur, especially at a small scale, where a restaurant, hotel or lodge may be family-run. But even at a large scale, expectations can be and are broken. Increasingly, for instance, men are taking on the job of contract cleaning for large businesses, and this may reflect their changing expectations in a job market which has been restructured for free trade – see Chapter 2 – and offers them few other opportunities.

As potential tourists, tour operators and service providers, can we expect that tourism businesses which claim to be responsible, sustainable and ethical will attempt to break these established roles? Or are we expecting too much of our responsibility and ethics that they should break down the traditional social, cultural and economic barriers? In many examples of new forms of tourism promoted by NGOs and INGOs in the LAC countries known to the authors, it is rare to find positions and roles which differ from the expected. It is interesting to note, however, that in many examples of new forms of tourism promoted by individuals or families and run primarily as a means of subsistence, it is reasonably commonplace to find men doing what were traditionally seen as women's jobs and vice versa. This difference may reflect the attempt of some new forms of tourism to offer tourism opportunities simply as a means of diversifying their sources of income whilst earlier forms of tourism have a clearer aim of profit maximisation. This is mere speculation based on anecdotal observation, but cannot be dismissed for that. In the absence of appropriate research, we accept that such differences may exist between different types of tourism opportunities.

Arguably more important than the changing of gender expectations is the changing of family expectations, also of course related to the issue of gender. In a society where poverty is on the increase (see Chapter 3), in which opportunities for economic earnings are few, and in which societal and economic pressures on the family cause unprecedented levels of family divisions and dysfunctions, it should come as no surprise that families send out on to the streets those members who have time on their hands or who have no other viable economic opportunities available to them in order to supplement the family income. This can involve begging, selling goods at traffic intersections, collecting and recycling waste items, and prostitution. None of these activities necessarily involves tourists, but it is obvious (from both intuition and experience) that they tend to occur in locations frequented by people with money. This group includes tourists, and exploitation of the sex of local people in tourist destinations by tourists often reinforces the attitudes we have discussed in this section; and it is the phenomenon of sex tourism that we address in the following section.

Sex tourism – a reflection of power

White males from First World countries have been travelling to Third World countries for many years and the sexual exploitation of women in this context is not new. But the recent phenomenon of sex tourism has emerged as a major activity in a number of countries. Edward Said depressingly catalogues the 'formidable structure of cultural domination' in his study of *Orientalism*.[11] The most persistent value of western cultural life is sex: sexual experimentation and fantasy, promise, desire, delight and unlimited sensuality. It is this key feature of Orientalism (or otherness) that has been so cleverly refined in the world of mass communications, but remains a standardised cultural stereotyping of 'the mysterious Orient' – smiling, servile and sexy. Sanchez Taylor makes it quite clear that Said's conclusions are just as applicable to other parts of the world as they are to the Orient: 'there is a long history of sexual exploitation of women under colonial rule and western men have long projected racist fantasies onto the "primitive"/natural Other. But the long-haul tourist industry is turning this kind of post-colonial fantasy into an item of mass consumption.'[12] And as anthropologist Franck Michel explains:

This is how an ordinary tourist, far from home, can end up doing the unthinkable. There is a more ready desire for self-transformation because the experience of travel allows tourists, organised or not, to divest themselves of any sense of responsibility. For the organised tourist, the Other, the native of colonial times, is there simply to serve and to be explained.[13]

Until the final few years of the twentieth century, organised sex tourism was most commonly perceived as being associated with the region of south-east Asia and in particular with the country of Thailand. Only in the last decade or two, with the growth of travel from the First World to the Third World, has the potential for this type of exploitation been realised by substantial numbers of men (sometimes referred to as 'sex-pats') who wish to partake of it. Moreover, the expansion of new forms of communication (such as the internet) has allowed organisation of the activity on a global scale. It is argued by some that sex tourism has become an almost ubiquitous feature of new tourism developments. The LAC countries have been no exception in this sense, and, to some degree, as the Thai authorities have half-heartedly tried to clean up their act, some part of the global focus of attention has shifted to Latin America and the Caribbean. Indeed, Patricia Campos of Cedeca, a human rights group in north-east Brazil, has suggested that 'UNICEF and international groups have really had their eyes trained on Asia', where sex tourism has long been a problem; and, because of this, 'there's been a migration of tourists of this kind to the north-east of Brazil – to Fortaleza, Natal, Recife. With the recent [tsunami] catastrophe in Asia, we're afraid that this shift will intensify.'[14]

If the extent of the sexual exploitation of adult women in a country is an indicator of the depth of poverty and misery in that country, then the incidence of sexual exploitation of its children is an indication that such poverty and misery has reached the extremes. As the Indian writer Shamsur Rabb Khan declares, 'sexual abuse of children is the most abominable violation of human rights. In fact, no crime can be as ghastly as using innocent children for sexual pleasure.'[15] Again, we have to point out that children have been abused since time immemorial, but the scale and systematic organisation of this abuse is a recent feature of modern society. In 1993, Professor Vitit Muntarbhorn, the United Nations Special Rapporteur on the sale and trafficking of children, stressed that such exploitation clearly occurs because of a demand created by tourists. Muntarbhorn

> observed that most current analysis of causes is too simplistic, focusing as it does on rural poverty and the sale of children for debt as the primary causal factor. He put forward the view that causes are related to 'poverty plus' other factors, and emphasised a need to look at other root causes, both local and transnational. In particular, he criticised the prevailing values of a consumerist 'hedenocracy' which dominate models of tourism and travel, and which gives consumerism priority over human rights.[16]

Throughout most of the 1990s, the Costa Rican government's newspaper for tourists included advertisements for sea fishing tours and other adventure tours as 'escorted' with the clear implication that the escort was there to satisfy the sexual desires of the predominantly US males who participated in such tours. That such advertisements should appear in a publication produced by the government of a country whose high-profile tourism reputation had been based on the promotion of its world of nature is surprising. In fact, such advertisements were still appearing in *Naturally Costa Rica*, a publication funded and endorsed by the Costa Rican Tourism Institute, the country's Chamber of Tourism and by CINDE, a Costa Rican government development agency, in the late 1990s. One of the persons responsible for the advertisements was Jorge Bubert Serrano who in 2001 was jailed for a minimal sentence of five years for his part as a ring leader of a network offering women and children for sex. He had also placed advertisements and brochures in hotels and newspapers promoting his business which had involved some of Costa Rica's five-star hotels.

It took some time for the exploitative nature of this activity to become fully apparent to many people in Costa Rica. Eventually, the opposition to such a trend was galvanised into pressuring the Costa Rican government to officially outlaw such advertisements. In 1998, however, the director of the Costa Rican Judicial Investigation Unit estimated that at least one half of one per cent of the close to one million foreign visitors to Costa Rica each year were so-called sex tourists. In other words, at least 5,000 of these tourists were sexual exploiters of women and children. And in 1999 a report by the Commission for the Defence of Human Rights in Central America (CODEHUCA) noted the absence of both legislation and political will to stop such exploitation.[17] Such practices had clearly not ceased in Costa Rica, but by 2003 at least they were no longer directly promoted by the national government. Even as recently as July 2003, *Mesoamerica*, a monthly publication of the Institute for Central American Studies (ICAS), reported that:

> Costa Rica has become a hot-spot for sexual exploitation of minors as demonstrated by the seventy internet sites that promote the country as a paradise for sex tourism. Some of the internet sites give advice on avoiding authorities and offer 'all-inclusive sexual holiday packages'. Reports state that an estimated 62 per cent of sexually exploited girls are victims of foreigners.[18]

We should be aware that this statistic of 62 per cent implies of course that 38 per cent are not victims of foreigners, which in turn implies that local nationals (not necessarily participating in the tourism industry) are also implicated in the activity.

The proportion of one half of one per cent of all tourists to the country of Costa Rica, as given above, is clearly only minor relative to the total flow of international tourists to the country. But if this proportion is repeated for many other LAC countries, the total represents a substantial number of people and a substantial amount of money earned by respected suppliers of international tourism such as the hotels in which they stay and the airlines which fly them there. Of course, it may not be appropriate to apply such a proportion to many or even any of the other LAC countries, and Costa Rica appears to have gained itself a significant reputation for sex tourism. But, lacking other indicators of the scale of the activity, we apply in

Table 7.1 An estimate of the annual number of sex tourists to LAC countries

Region	International tourist arrivals	0.5% – sex tourists
Caribbean	17,334,000	86,670
Central America	4,270,000	21,350
South America	15,432,000	77,160
Total	37,036,000	185,180

Source: World Tourism Organisation (2001) *Tourism Market Trends: Americas – 2001*. Madrid: WTO.

Table 7.2 An estimate of the annual commercial value of sex tourism

Region	International tourism receipts (US$ millions)	0.5% – sex tourism
Caribbean	17,158	85.79
Central America	2,926	14.63
South America	12,122	60.61
Total	32,206	161.03

Source: World Tourism Organisation (2001) *Tourism Market Trends: Americas – 2001*. Madrid: WTO.

Tables 7.1 and 7.2 the proportion of 0.5 per cent to the whole region under consideration. The tables suggest that a total of 185,180 of the 37 million annual tourists to the LAC countries may be sex tourists and that this sector of the industry can certainly be described as a multi-million-dollar earner.

Other indications of the extent of the sexual exploitation of children do exist, but they are no more than tenuous indications. In an unreferenced work Barger Hannum, for instance, cites a 1994 estimate by ECPAT (Ending Child Prostitution, Pornography and Traffic) of 500,000 children in Brazil involved in the sex industry and a Colombian Ministry of Justice report of at least 25,000 child prostitutes in that country.[19] She also cites a Casa Alianza (see below) estimate of 5,000 Honduran street children involved in sex tourism and suggests that the same problems are found in Paraguay, the Dominican Republic and Venezuela. Regarding Cuba, Martha Honey cites the fact that the number of Italian tourists, 'many of them pursuing "sexual tourism", jumped almost eightfold between 1991 and 1996, from some 24,000 to more than 192,000, according to the Ministry of Tourism'.[20]

One NGO pursuing the problem of sexual exploitation of minors in Latin America is Casa Alianza, whose head office is in San José, the capital of Costa Rica. Whilst Casa Alianza's usual activities include political and legal lobbying in all of the countries in which it works (Costa Rica, Guatemala, Honduras, Mexico and Nicaragua), its programmes most commonly involve the giving of care and advocacy for street children. These are programmes which attempt to wean street kids off the street and all its attendant dangerous activities and to give them a home that represents love and security and an alternative to a life of crime and victimisation. In Costa Rica, however, the organisation serves principally as a lobby group against the illegal exploitation of minors for the purposes of sex which happens usually to be practised by white, middle-aged or old males from the USA and Europe. The organisation has denounced over 400 cases of commercial sexual exploitation of children in Costa Rica, but because of insufficient funds the authorities are unable to address the problem and to bring to justice the criminals involved.

Costa Rica is not the only Latin or Caribbean country to gain a reputation for sex tourism. According to Casa Alianza, the same type of networks now appear to operate in other Central American countries, and South American and Caribbean countries are not without their own cases of sexual exploitation. The international trafficking of children for sexual purposes increased immediately after Hurricane Mitch hit Central America in 1998. In his 1999 address in Switzerland to the 24th session of the United Nations Working Group on Contemporary Forms of Slavery, Bruce Harris (former Executive Director of Casa Alianza) gave an idea of how sexual exploitation affects children of the region. Extracts from his account are given in Box 7.1.

Box 7.1   Central American children's route into sex tourism

The situation of children in the region has recently been made worse by the horrendous impact of Hurricane Mitch in Central America . . ., especially in Honduras and Nicaragua. Now waves of children and teenagers are fleeing these countries and heading north using any means they can in order to reach the supposed dreamland of the United States of America . . . Behind the train station in the Mexican border town of Ciudad Hidalgo, Chiapas, just a few kilometers from the Guatemalan border, are the brothels that entertain the scores of truck drivers waiting to cross south into Guatemala. Most of the girls being exploited there are barely 13 or 14 years old and most are from El Salvador, Honduras and Nicaragua. In their desperation for money to head north, they sell their bodies to make less than five dollars a trick.

For the children who stay behind in their own countries, more and more are falling into the grips of prostitution in order to be able to survive. In the area around El Belén in Tegucigalpa, the cheap, run-down bars offer pre-pubescent girls for unprotected sex for less than three dollars. No condoms, no safe sex. With 40,000 estimated cases, Honduras has the highest HIV rate in all of Central America. Most of these people are developing full-blown AIDS between the ages of 21 and 29, which means that most are being infected between the ages of 12 and 18.

Source: Bruce Harris, (1999) Presentation to the 24th Session of the United Nations Working Group on Contemporary Forms of Slavery. Geneva, Switzerland, June.

As the first part of the passage in Box 7.1 makes clear, prostitution services and the exploitation of children are not solely associated with the international tourism trade. Harris identifies truck drivers as major clients, and the wealth of work from Casa Alianza on this issue makes it evident that much of the exploitation is by national clients. Indeed, it is likely that most of the prostitution which occurs around the border zone of northern Mexico is aimed at locals, and the Tegucigalpa case cited in Box 7.1 is patronised largely by urban Hondurans rather than international tourists. Harris himself has acknowledged that 'despite sex tourism being a major problem, the sexual abuse of children is dominated by local people'.[21]

Both of the cases in Box 7.1 highlight the association between prostitution and migration borne out of desperation. In the first case, the girls are attempting to earn their way into the 'promised land' of the United States where economic opportunities are greater than in their own country. In the second case, El Belén is the destination for girls migrating from poverty within the same country (Honduras), whether rural or urban.

This migration is a reflection of the differential outcomes of the economic and political processes operating in the LAC countries. On the one hand, current neoliberal policies – see Chapter 2 – have created a spiral of poverty and desperation for large sectors of the population, leaving them devoid of opportunities in their home environment. This creates the economic imperative of survival and forces many either to move to find such opportunities or to create for themselves illegal opportunities such as prostitution. The few legal economic opportunities which do arise through such policies are increasingly found in *maquilas*, the sweat-shop factories most commonly found in Free Trade Zones and particularly associated with the employment of large numbers of young women and often located in zones or points of transit. Biemann (interviewed by Szeman) has noted that 'large-scale prostitution activities have settled around Free Trade Zones all over the world'.[22] Biemann and Szeman relate this directly to the impossibly low level of pay in the *maquilas* which 'forces many of them to earn a supplementary income on weekends through prostitution'. They explain the link between neoliberal economic policies and the incidence of this 'sexualisation' (as they call it) of the female *maquila* workers thus:

> it is on a corporate level since transnationals benefit directly from getting labour for pocket money by making women dependent on commodifying their bodies. The interpenetration of the industrial market and the sexual market is not an interesting side effect, it is a structural part of global capitalism.[23]

The US–Mexico border zones around Ciudad Juárez and Tijuana are referred to as '*zonas de tolerancia*' (zones of tolerance) where the presence of a large number of very poorly paid young women (working in the *maquilas*) creates a supply of prostitutes to serve the US and Mexican markets for sex. The case of Ciudad Juárez in Mexico is given in Box 7.2, which helps to explain the powerlessness behind the actions of the victims of this form of exploitation.

Box 7.2   Tourism and sex in Ciudad Juárez

The pattern which dictates the arrangement of the tourist district is quite similar to the one which dictates the arrangement of the red light districts, known in Spanish as *zonas de tolerancia*, or more simply, *zonas*. These *zonas* tend to emerge toward the fringe of the tourist district, in the middle of the old urban core . . . The facilities in these districts provide music, dancing, floor shows, room accommodations, and food service. There are few real brothels to speak of, but as prostitution is not against the law in Mexico, many places accommodate it. Also, within the *zona* there are bars and clubs where prostitution does not take place. Though one may suspect that since these places are on the periphery of the tourist district that the zones are catering to foreigners, in reality, studies have suggested that up to 70 per cent of prostitutes service only Mexican clients (Arreola and Curtis, 108–109).

. . . the *zona* within Ciudad Juárez is . . . about the size of ten city blocks. The average clientele may also have a higher percentage of Americans than in other cities . . . The transportation within the city and the business of prostitution are linked in by a separate sub-culture which exists among taxi drivers, club owners and prostitutes. The owners established relationships and deals with the taxi drivers in order to keep a steady supply of clients coming through the door (Martínez, 177).

(Fay and Vender 2003)

References:
Fay, R. and Vender, J. (2003) 'Cultural Landscape of Ciudad Juárez', http://web. scholars.psu.edu/~juarez/drafts/fay.rtf.
Arreola, D. and Curtis, J. (1993) *The Mexican Border Cities: Landscape Anatomy and Place Personality*, Tucson: The University of Arizona Press.
Martínez, O. (1994) *Border People – Life and Society in the US–Mexico Border-lands*, Tucson: University of Arizona Press.

Laurence Gray of World Vision has also spoken of the commodification of children and describes the global tourism industry as promoting:

the idea that you can go somewhere and use your spending power to have a great experience and enjoy yourself – and if your pleasure is to go with children, then that can be legitimised in some way because it's part of the experience. It's a power relationship, too, between tourists and sex workers. It's not a meeting of equals. It's a meeting of someone with dollars with someone in great need.[24]

The powerlessness of the women and children who prostitute themselves is further evidenced by Polly Pattullo's description of them (in the Dominican Republic) as 'usually impoverished young women with children who turn to prostitution from economic need'.[25] Pattullo suggests that 'the Dominican Republic has now joined Thailand and the Philippines on the list of countries known for its exploitative sex industries'. She also cites Cuba as a country to which the sex industry has returned alongside the return of the tourism industry: 'The eradication of prostitution was one of Castro's proudest boasts. Now, with more tourists and more hardship, sex tourism is back.' In Cuba, however, at least officially, prostitution is not seen as a result of economic desperation, but more a result of moral laxity. (It should be noted that this official view may not be shared by many of the participants in the Cuban sex industry.) An empathy with the genuine Cuban attempt to integrate morality with politics, as opposed to the hypocritical stance of western governments, should not lead us into a denial that an analysis of the structure of power is essential to the understanding of the nature and dynamics of the sex tourism industry. Power, or lack of it, lies behind all of the roles of the actors in the industry, whether it is expressed politically, physically or psychologically.

The other face of this powerlessness is the power exercised by the exploiters, be they the traffickers, the internet ring organisers or those who consume the service. Bruce Harris makes clear his view that this power is political as well as physical: 'The fact that children are being exploited in every shape and form imaginable, is a political decision.'[26] Here he is alluding to the ability (or otherwise) of governments to choose to legislate against such exploitation and to choose to enforce such legislation. Third World governments are often 'excused' such 'luxuries' as child protection laws on the grounds that as emerging countries they do not have an adequate legal infrastructure and that they have more urgent and crucial matters to deal with, such as economic and development policies. It is difficult to imagine anything more urgent or crucial than the care of the country's future, its children, which only serves to highlight the misguided assumption behind the excuse.

That it is a political decision reflects the power of the First World over the Third, the power of males over females, the power of age over youth and the power of rich over poor. But the exercise of this power is not solely for the purpose of the selfish, physical gratification of the male. As Franck Michel states,

> The main reasons for the unprecedented growth in mass sexual tourism include worsening poverty; the liberalisation of sexual markets, which encourages trafficking for prostitution; the persistence of patriarchal, sexist societies; and the degradation of the image of women through widespread, normalised sexual violence. There is also the explosive growth of international tourism and migrancy stimulated by two special factors: the democratisation of travel (huge numbers of tourists flying cheaply everywhere) and the hypersexualisation of the young, fostered by the media obsession with sexual violence.[27]

Making the same link between politics, patriarchy and sex, Vanessa Baird believes that 'autonomy – especially for women – is a threat to authoritarian and patriarchal control'.[28] In reinforcing the powerlessness of women in the minds of both men and women, sexual subjugation of women therefore serves a political end in bolstering the prevailing political and economic order.

Such exploitation does not go unchallenged. The organisation ECPAT used to stand for End Child Prostitution in Asia Tourism and campaigned specifically against child sex exploitation in the countries of south-east Asia. More recently, as awareness of the problem and its global spread has grown, it changed its name to Ending Child Prostitution, Pornography and Trafficking of Children for Sexual Purposes, using the same initials, and now has offices in 45 countries (including some of the LAC countries) and addresses child exploitation worldwide. In 1998 ECPAT produced a code of conduct for tour operators in relation to the commercial sexual exploitation of children in travel and tourism. The code of conduct lays emphasis particularly on staff training, company policy and awareness raising among customers and the public in general. It was developed in collaboration with TUI Nordic, a major European tour operator. Partly as a result of ECPAT's campaigns and efforts, the organisation has been able to report that the number of paedophiles in Thailand is decreasing. But this decrease has come at the expense of other countries, especially in Central America, where, in collaboration with regional NGOs, ECPAT also now operates.

In most of the Central American countries, the organisation Casa Alianza attempts to defend the rights of street children. It is the Latin American branch of the New York-based Covenant House, a leading advocate of children's rights. It campaigns especially vigorously against the sexual abuse of minors and their trafficking for the purposes of sexual exploitation and has exposed many such cases which have typically involved middle-aged or old-aged males from North America or Europe as the traffickers or abusers. Casa Alianza has widely publicised the links between this exploitation and tourism and in particular has stressed the need for the tourism sector to commit itself to the fight against the commercial sexual exploitation of children.[29] As it states, 'The supply will not decrease until the demand is dealt with'.[30]

Laying emphasis, however, on child sex tourism may lead to the danger of excusing or absolving the sex tourists' sexual abuse of adults (male or female). In this sense, the distinction between tourists' sexual abuse of children or adults is spurious, and it is important that in focusing on the horror of child sex abuse we do not turn a blind eye to traditional prostitution.[31]

Women sex tourists

Sex tourism is not always the practice and preserve of white and wealthy males from prosperous countries. In recent years a number of studies have highlighted the phenomenon of women sex tourists (largely, but not exclusively, from the First World) exploiting males in Third World countries.[32] As Julie Bindel suggests, 'we can be fooled into thinking that the freedom [of women] to behave like men is liberation'.[33] We realise here that the First World/Third World dichotomy is too simplistic and tends to hide the complexity of movements in other directions. Its simplicity, however, does not render it irrelevant or entirely inaccurate; and so we choose to retain the terms, bearing in mind our earlier justification – see Chapter 1 – for their use.

In 2003 Julie Bindel reported on work in Jamaica, specifically at the Negril beach area which attracts a high proportion of Jamaica's 1.3 million tourists every year, primarily from North America and Europe. She pointed out that:

> There are some obvious differences between female and male sex tourism. Although the beach boys are objectified and often sexually humiliated by the female tourists, they tend not to fear or experience violence and sexual aggression, such as being beaten and raped, in the way that female prostitutes routinely do. Nor are they vulnerable to criminalisation, unlike female prostitutes whose activities are illegal.[34]

The differences, however, go beyond the physical and the legal aspects of the trade. They include the perceptions held by all parties involved in the business. Bindel cites Sanchez Taylor and O'Connell Davidson's findings from work in Negril, Jamaica, and in the Dominican Republic, where over 240 women holidaymakers were interviewed. Of the third who admitted to some form of sexual relationship during the holiday, '60 per cent admitted to certain "economic elements" to their liaisons, [though] they did not perceive their sexual encounters as prostitute-client transactions, nor did they view their sexual partners as prostitutes'.[35] Sey reinforces the idea that 'These women rarely consider themselves sex tourists'.[36] This difference in perceptions is also marked for the male prostitutes themselves. Bindel reports that 'Not one of the men I spoke to would admit to money being a prime motivation for their liaisons'.[37] Sanchez Taylor and O'Connell Davidson consider that this perception of the difference between male and female sex tourism is also generally held by researchers and theorists.

According to Julie Bindel, the Sanchez Taylor and O'Connell Davidson research suggests that: 'the reason many female tourists are able to delude themselves into believing they are not prostitute users lies in their racialised power over the men'.[38] This is a reference to the myth (if that is what it is) that black men are physically well-endowed and sexually hyperactive. With the help of this belief, the female sex tourist can persuade herself that her gigolo is interested in her for reasons other than finance. But, as Sey points out, the women maintain a power over the relationship through money and status.[39] The women visitors have money; by and large the local men do not. Indeed, many local men pursue this activity precisely because there are no or few other economic opportunities available to them, and many hope to 'hit the jackpot' by gaining the friendship of a tourist who will help them get to Europe or North America.

The existence of this power relationship is supported by Pattullo, who describes the beachboy phenomenon as based upon the dependency of the hosts, whom she describes as:

> usually poorly educated and unemployed, [and who] offer their sexual services in return for money or temporary support to affluent, often middle-class, white tourists, whose age and marital status are pretty well irrelevant. While it might not be called prostitution, the contract has a barely disguised financial

basis, with the tourists providing meals, drinks, transport, money and clothes in exchange for sex.[40]

Despite this relationship of power, the economic pressures upon male prostitutes are normally less than those faced by female prostitutes. Whilst many of the former do earn an income for their families, the latter are more often the head of household and bear the responsibility for raising children. Female prostitutes are also much more likely to lose a proportion of their sexual earnings to a middleman or pimp.[41]

One feature of the industry associated with both male and female sex tourism is that its financial benefits are not received solely by its participants. Agents and elements of the tourism and travel industry themselves are major beneficiaries of the activity. As O'Connell Davidson describes them, the major beneficiaries 'are probably the airlines which transport prostitute users half-way around the globe, the hotels (many of which are owned by international conglomerates) in which they stay, and the travel agents which arrange their flights and accommodation'.[42] In 1995 the World Tourism Organisation issued a strong statement on the prevention of organised sex tourism with appeals to the travel trade to actively join in efforts to eliminate this aspect of the trade. The fact that sex tourism has shown no signs of abating since then perhaps suggests that rather more than a strongly worded statement is required if the organisation's members are to take effective action.

We have already raised the issue of whether tourism brings development or development brings tourism – see Chapters 2 and 3. An increasing incidence of sex tourism in a locality may in some minds suggest a lack of or loss of development rather than a mark of the progress which development is supposed to imply. In the context of sex tourism, it is tempting to suggest that, whilst sex tourism is nothing new, the current scale of its organisation now seems to have extended the activity beyond the old-style colonialists and the ex-pat community to include potentially all international tourists. And this level of organisation appears to have arisen alongside the rate of growth of the tourism industry. Without offering a judgement, it seems to us that prevailing opinion would not consider the growth in sex tourism as representing what is popularly perceived as 'development'.

The recent growth in sex tourism has occurred alongside a growth in the incidence of sexually transmitted diseases. For tourists the health hazards of casual and/or paid sex represent another significant issue for consideration within any analysis of responsibility in the tourism industry; and it is to this issue that we turn next.

Sexual health and sex education

Plentiful and promiscuous sex has the awkward tendency of leading to sexually transmitted diseases (STDs), which for the tourist represent one of the two major disadvantages of the activity, the other being the possibility of being discovered and convicted for illegal activities. As is commonly known, the use of condoms is an important means of preventing the transmission of sexual diseases, and the crucial importance of their use can be and normally is promoted through education. The United Nations Population Fund (UNFPA) has made prevention the centrepiece of its fight against STDs, especially HIV/AIDS. 'Prevention includes promoting safer sexual behaviour among young people, making sure condoms are readily available and widely and correctly used, empowering women to protect themselves and their children, and encouraging men to make a difference.'[43] The use of condoms, however, is generally perceived, especially by men, as something that detracts from the experience of sex. Indeed, many men are prepared to pay more for their time with a prostitute if they are allowed unprotected vaginal penetration. Often the power of money, and extra money, is greater than the power of education. A number of short case studies from LAC countries (given in Box 7.3) illustrate this problem and the significance of effective sex education.

Box 7.3   Evidence of sexual health problems

Catalina [from the Dominican Republic] does not have lodgings of her own but accompanies the tourists and expatriates who pick her up anywhere they choose to take her. She describes some of her clients as 'easy' . . . Others are more demanding, and she has experienced violence at the hands of some clients. But perhaps most telling of her powerlessness is her inability to enforce condom use, despite her knowledge of AIDS and how it is contracted.

Catalina has a nine-month-old child fathered by a sex tourist (who contributes nothing towards the baby's maintenance), as she never used condoms when she started to prostitute. Catalina still sometimes acquiesces to clients' demands for unprotected vaginal penetration because 'they offer you a lot of money for sex without a condom'.

(O'Connell Davidson, 1999: 8)

Of 143 homeless children interviewed, 100 per cent had at least one sexually transmitted disease, and 48.1 per cent had been sexually abused by a member of their family. (Casa Alianza study, 'Former Honduran Street Girls Travel to Canada for International Summit of Sexually Exploited Youth', 2 March 1998.)

(CATW Factbook – Honduras, 2003)

The increase in HIV infection in Trinidad and Tobago is attributed to sex tourism, prostitution and pornography.

(CATW Factbook – Trinidad and Tobago, 2003)

Note:
CATW: Coalition Against Trafficking in Women.

References:
O'Connell Davidson, J. (1999) 'Doing the Hustle', *Tourism in Focus* 30, Winter 1998/9. London: Tourism Concern.
CATW (2003) CATW Country Factbooks, www.catwinternational.org/fb/.

AVERT is an international HIV and AIDS charity based in the UK which aims to avert HIV and AIDS worldwide. Through its website[44] the organisation claims to take education and information on the issues associated with STDs to almost every country in the world. AVERT promotes sex education as the development of general life skills such as 'negotiation, decision-making, assertion and listening'. It is unequivocal in confronting the 'just-say-no' lobby's 'attempts to impose narrow moralistic views about sex and sexuality on young people', which it describes as having failed.

AVERT also enters the debates around the origins, existence and prevalence of AIDS and the links between HIV and AIDS with a vigorous and well documented defence of the reality of the threat that HIV/AIDS poses to human populations. The fact that it feels it necessary to make such a vigorous defence is a pointer to the reader that the world of medical science has not always been entirely in agreement about these issues. Critics of the general consensus suggest that the problem of AIDS (auto-immune deficiency syndrome) has been wildly over-stated. Stockwell, for instance, cites substantial evidence from the USA and Africa which suggests that medical science went on holiday whilst the media, specific individuals, the drug companies and government health departments claimed that AIDS would be an epidemic of biblical proportions.[45] Specifically, Stockwell lambasts the pushing of the AZT drug as a solution to AIDS, and claims that this has itself caused many deaths. In Africa he (and others) accuse western medicine of incorrectly diagnosing AIDS and of attributing many deaths incorrectly to AIDS.

The science behind the criticism is more clearly represented by the work of Peter Duesberg[46] who believes that the AIDS epidemic in Third World countries can be distinguished neither epidemiologically nor clinically from conventional diseases and mortality. (The relevant part of Duesberg's work refers specifically to African AIDS, but here we have extended it to all Third World countries because it is based on the assumption that many deaths attributed to AIDS would more correctly be attributed to diseases associated with poverty and malnutrition. Duesberg also hypothesises that American and European 'AIDS diseases are brought on by the long-term consumption of recreational drugs and/or AZT itself, which is prescribed to prevent or treat AIDS'.) Moreover, Duesberg questions the epidemiological statistics behind the warnings of catastrophe which informed then President Clinton's designation of AIDS as a 'threat to US national security . . . spurred by US intelligence reports that . . . projected that a quarter of southern Africa's population is likely to die of AIDS'.[47] Duesberg accuses the World Health Organisation (WHO) and the American AIDS establishment of propagating a 'deceptive AIDS propaganda [that] biases a scientific analysis of African AIDS by all those who are not aware of the facts'.

By raising this issue in this section we are not expressing our support for one side of the argument or another. It should be noted that Duesberg is essentially alone in these views against virtually all other medical professionals, amongst whom there is clearly a widespread belief that AIDS is a major killer disease, and the fact that the South African government has come to accept the prevailing wisdom on the issue now considerably reduces the strength of the dissenting viewpoint.[48] There is almost universally accepted evidence that it results either from unprotected sex or by contact with HIV-positive blood. These beliefs are not being questioned here. It is our intention simply to note that there exists some dissent from the generally accepted wisdom of the day and to refer the reader to other works[49] in order to take study of the issue further. It is of interest and relevance to note, however, that these differences of opinion regarding the nature and prevalence of AIDS are reflected in different approaches to education on the topic. In 2003 Brazil and the USA, for example, introduced and implemented HIV prevention strategies with sharply divergent styles. According to Ballvé:

> politics and ideology account for the differences . . . The two countries' approaches to HIV/AIDS differ not only in content but also in communications strategies . . . Brazil combines sophisticated mass marketing with the force of its federal communication apparatus to give HIV prevention mass appeal in a diverse nation . . . the US Centers for Disease Control (CDC) . . . says that it is best to give local governments and community-based groups control over HIV prevention campaigns.[50]

US HIV prevention programmes have a tendency to focus on high-risk groups. In Brazil, on the other hand, Arthur Raichman, head of HIV programmes in São Paulo state, speaks of an 'epidemic of prejudice' that fuels global HIV. 'I think this concept of a risk group helped create this epidemic,' he says.[51]

Referring to Haiti, Polly Pattullo suggests that sex for the tourist used to be a real bargain, that is, before 'Aids became the ultimate deterrent to the sex industry'.[52] That tourists immediately stopped visiting Haiti in the mid-1980s as a result of scare stories about the then new and powerful disease, AIDS, cannot be denied; although whether there was any truth in the scare stories (as regards the disease's prevalence in Haiti) is much more open to debate as other, political and security factors may well have been partially responsible for the dramatic drop in visiting tourists.

The question arises of whether AIDS really is the ultimate deterrent to the sex industry. If so, it might be worth reinterpreting Prosser's 'tidal wave of the pleasure periphery'[53] as a tidal wave of sexual pleasure periphery. At least, that might be a suitable title if it were not for the fact that for many involved the phenomenon of sex tourism does not represent pleasure. A more suitable term might therefore be the 'tidal wave of sexual exploitation periphery'. Referring to the global spread of tourism, Prosser identifies five peripheral regions of the world which have been successfully and historically commodified for the tourism industry over the last century. In the reinterpretation for the sex tourism industry, assuming that the first region is the origin region of western Europe and North America, the remaining four regions would be:

- pre-twentieth-century European colonies (anywhere in the formerly colonised world)
- south-east Asia and Africa before AIDS

* the Caribbean Islands
* Latin America.

Prosser's idea refers to 'an admired elite' who create a fashion to which others of lesser socio-economic status later aspire, or in tourism terms wish to emulate. As the fashion takes hold of much greater numbers, the 'admired elite', driven by the desire for novelty, uniqueness and exclusivity of experience, then move on to a new region in the pleasure periphery. Reinterpreted to apply to the sex tourism industry, the exploiter 'elite' discover a region unlikely to be visited by others and unlikely to be dogged by problems of STDs. As awareness and popularity of this region increase, exclusivity is lost and sexual diseases multiply, the exploiter 'elite' moves on to another region, not known for its sex tourism. As the process continues, the exploiter 'elite' moves on in search of new destinations where the incidence of sexually transmitted diseases and, particularly recently AIDS, is nil or low. Perhaps this same process helps to explain why child sex tourism also appears to be an increasing phenomenon.

Speaking of Managua, the capital city of Nicaragua, Zambrana quotes Luis Cuadra, the Director of the organisation SILAIS (Local System of Integrated Health Care): 'The number of cases of AIDS has increased with the increase in the number of tourists to the city. The problem which we are facing is that neither the girls nor the foreigners are using the necessary protection to avoid the contagion of AIDS.'[54] Referring to the same country, Romero makes an interesting comparison between the situation in Nicaragua and other LAC countries by citing the Nicaraguan Women's Network Against Violence:

> if tourism is going to become an economic alternative in the country, the government has to protect its children, women and youth, because they don't want to occur what happens in Costa Rica, Cuba and the Dominican Republic, where tourism has become an alternative engine of economic growth, but where also there has been a worsening situation of sex tourism and sexual exploitation of minors.[55]

Table 7.3 gives the most recent figures on the incidence of HIV/AIDS from the AVERT website, which describes the Caribbean as 'the second-most affected region in the world'[56] (after sub-Saharan Africa). AIDS is now one of the leading causes of death in the Caribbean countries, and Haiti stands out as the worst case. 'The predominant route of HIV transmission in the Caribbean is heterosexual contact. Much of this transmission is associated with commercial sex, but the virus is also spreading in the general population.' Regarding South America, however, Avert suggests that in most countries 'HIV is not generalised but is highly concentrated in populations at particular risk'. Despite many constraints, however, the organisation praises the region for the admirable progress it is making in the provision of treatment and care:

> Brazil in particular is now producing AIDS drugs at a fraction of the cost of the big multi-nationals . . . Other countries with high levels of access to

Table 7.3 Estimated number of people living with HIV/AIDS at end 2005, and estimated number of deaths due to AIDS during 2005, LAC countries

Country	Population (millions) (2003 est.)	Living with HIV/AIDS	Adult (15–49) rate (%)	Deaths due to AIDS
Argentina	38.4	130,000	0.6	4,300
Belize*	0.3	3,700	2.5	<500
Bolivia	8.8	7,000	0.1	<500
Brazil	178.5	620,000	0.5	14,000
Chile	15.8	28,000	0.3	<500
Colombia	44.2	160,000	0.6	8,200
Costa Rica	4.2	7,400	0.3	<100
Ecuador	13.0	23,000	0.3	1,600
El Salvador	6.5	36,000	0.9	2,500
Guatemala	12.3	61,000	0.9	2,700
Guyana*	0.8	12,000	2.4	1,200
Honduras	6.9	63,000	1.5	3,700
Mexico	103.5	180,000	0.3	6,200
Nicaragua	5.5	7,300	0.2	<500
Panama	3.1	17,000	0.9	<1,000
Paraguay	5.9	13,000	0.4	<500
Peru	27.2	93,000	0.6	5,600
Suriname*	0.4	5,200	1.9	<500
Uruguay	3.4	9,600	0.5	<500
Venezuela	25.7	110,000	0.7	6,100
Total: Latin America	504.6	1,600,000	0.5	59,000
Bahamas*	0.3	6,800	3.3	<500
Barbados*	0.3	2,700	1.5	<500
Cuba	11.3	4,800	0.1	<500
Dominican Republic	8.7	66,000	1.1	6,700
Haiti	8.3	190,000	3.8	16,000
Jamaica	2.7	25,000	1.5	1,300
Trinidad and Tobago	1.3	27,000	2.6	1,900
Total: Caribbean	38.7	330,000	1.6	27,000

Source (column 2):
UNFPA State of World Population: Demographic, Social and Economic Indicators, 2004.
Source (columns 3, 4 and 5):
AVERT website, www.avert.org which takes its data from UNAIDS/WHO reports (2004) on the global AIDS epidemic.

* Population data for these countries from the Population Reference Bureau website: www.prb.org

antiretroviral drugs include Argentina, Chile, Mexico, Uruguay and Venezuela. However, in Ecuador, Paraguay and the poorer countries of Central America a large proportion of people are still unable to access treatment.[57]

Regardless of the viewpoint adopted on the issues of AIDS causation and treatment, there is no dissent from the fact that there exists a wide range of sexually transmitted diseases, and that some of these threaten life. It is crucial therefore to examine the nature of the problems of sexual health and to acknowledge the importance of education in sexual health for all involved in the activity.

In 1995, the WTO/OMT issued a strongly worded statement on the prevention of organised sex tourism, stressing in each sentence the significance of the role of education of all parties concerned, and acknowledging, by implication, the role that tourism has played in the sexual exploitation of women and children. Its requests of governments of both tourist sending and receiving countries are given in Box 7.4, and its appeals to the travel trade are given in Box 7.5.[58]

The WTO/OMT's statement and guidelines lay emphasis on education, whether it be of the tourists or tourism service providers. More recently President George W. Bush, addressing the General Assembly of the United Nations in September 2003, likened the trafficking of individuals for sexual purposes to the slave trade and pledged resources to combat it and educate against it. The statements from both the WTO/OMT and President Bush were strong, but the incidence of sex tourism since the WTO/OMT's statement has gone on rising and there are serious doubts whether the Bush commitment is really intended or well targeted. Under the

---

Box 7.4 WTO/OMT requests to governments regarding sex tourism

The WTO/OMT requests governments of both tourist sending and receiving countries to

Mobilise their competent departments, including National Tourism Administrations, to undertake measures against organised sex tourism;

Gather evidence of organised sex tourism and encourage education of concerned government officials and top executives in the tourism sector about the negative consequences of this activity;

Issue guidelines to the tourism sector insisting that it refrains from organising any forms of sex tourism, and from exploiting prostitution as a tourist attraction;

Establish and *enforce*, where applicable, legal and administrative measures to prevent and eradicate child sex tourism, in particular through bilateral agreements to facilitate, *inter alia*, the prosecution of tourists engaged in any unlawful sexual activity involving children and juveniles;

Assist intergovernmental and non-governmental organisations concerned in taking action against organised forms of sex tourism.

Source: WTO/OMT (1995) WTO Statement on the Prevention of Organised Sex Tourism. 11th General Assembly of the WTO/OMT at Cairo, Egypt, October.

Box 7.5   WTO/OMT appeals to the travel trade regarding sex tourism

The WTO/OMT appeals to the travel trade to

1   *Join efforts* and cooperate with non-governmental organisations to eliminate organised sex tourism, at both the origin and destination of travel flows, by identifying and focusing on the critical points at which this activity can proliferate;

2   *Educate staff* about the negative consequences of sex tourism, including its impact and the image of the tourism sector and tourist destinations, and invite staff to find ways to remove commercial sex services from the tourism offer;

3   *Develop and strengthen* professional codes of conduct and industry self-regulation mechanisms against the practice of sex tourism;

4   *Adopt* practical, promotional and commercial measures, such as, for example, positive self-identification of enterprises which refrain from engaging in sex tourism; banning commercial sex services, in particular involving children, on the contracted tourism premises; providing information to travellers about health risks of sex tourism, etc.;

5   *Warn* tourists particularly against engaging in child sex tourism, denouncing its criminal nature and the manner in which children are forced into prostitution;

6   *Encourage* the media to assist the tourism sector in its action to uncover, isolate, condemn and prevent all organised forms of sex tourism.

Source: WTO/OMT (1995) WTO Statement on the Prevention of Organised Sex Tourism. 11th General Assembly of the WTO/OMT at Cairo, Egypt, October.

---

headline 'Moral ties attached to US AIDS cash', Julian Borger reports that 'The conditions [imposed by the Bush Administration on funding for the fight against AIDS] reflect a push by conservative Christian groups for emphasis on sexual abstinence rather than precautions'.[59]

Human rights groups have pointed out that the problems of sexual exploitation and sex tourism will not be solved until the root causes of poverty and poor education are tackled. Jodi Jacobson of the Centre for Health and Gender Equity stated that 'The President has consistently undercut family planning and other reproductive health services desperately needed by women worldwide, including most recently by cutting off funds to six organisations working to prevent HIV infection among women in refugee settings'.[60] Writing in *The Guardian*, Nick Wadhams also pointed out that part of the President's proposal was to deny funding to 'any group or organisation working with female prostitutes that does not have a policy explicitly opposing prostitution'.[61] In July 2005, Brazil announced that 'it would refuse to accept $40 million (£22 million) in American aid rather than stigmatise prostitutes who Brazilian health workers said were essential to their anti-AIDS strategy'.[62]

The prejudice and denial of reality exhibited by President Bush has found support within the Catholic Church in the UK, where the organisation Catholic Action Group has initiated a boycott of funding for the Catholic international aid and development agency, CAFOD. CAFOD's sin is to suggest that condoms may be an acceptable way of combating the spread of AIDS in the developing world.[63]

Perhaps illustrating the breadth of opinion within the Catholic church, the UK-based Catholic organisation Progressio (formerly known as the Catholic Institute for International Relations) has been heavily critical of President Bush's AIDS Relief Plan, the key elements of which are *Abstinence, Be Faithful,* and *Condoms,* or the ABC strategy. Progressio claims that the emphasis is strongly on the A and B elements, reflecting the influence of the religious right. A preferable strategy, according to Progressio and most of its partner organisations in the LAC countries, would be preventive and based on *Condoms, Rights* and *Equity,* or the CRE strategy. The CRE strategy emphasises the importance of abolishing inequalities between men and women and of basing prevention on human rights. As Pablo Soto states in Progressio's journal,

> Neither the condom on its own, nor abstinence or fidelity together, will be able to slow the advance of the pandemic without this perspective. Because when we can all freely exercise our rights without discrimination or coercion of any kind, not only will the process towards obtaining the Millennium Development Goals relating to HIV and AIDS be accelerated, but other closely-linked goals will be met: reduction of poverty, raising the level of health and education, increasing productivity and raising the standard of living.[64]

So, on an issue which could be expected to attract a high level of accord and consensus such as the education of all parties involved in sex tourism, there is contention and discord. Notwithstanding this discord, in recent years there has been something of a corporate *putsch* to promote partnerships and educational projects covering issues associated with sex tourism. One such project has been sponsored by TUI Nordic of the TUI Group, a leading European tour operator, which has implemented ECPAT's code of conduct on the commercial sexual exploitation of children (CSEC). By doing so, it commits itself to train its staff to combat CSEC, provides its customers with information about CSEC, requires its suppliers to work against CSEC, includes appropriate policy statements about CSEC and reports annually on its achievements in this regard. In Brazil and the Dominican Republic, it has conducted pilot awareness raising programmes with its customers through brochures, its website, welcome meetings and in hotel books and information leaflets. An extract from TUI Nordic's hotel contract is included in Box 7.6. It is interesting to note that the company reports a degree of success in ECPAT's work seen in the decrease in the number of paedophiles visiting Thailand, but adds that:

Box 7.6   Extracts from TUI Nordic's hotel contract

The commercial sexual exploitation of children is an increasing problem. Specifically, there is a growing problem with child prostitution in tourist destination countries. TUI Nordic recognises that the problem is linked with tourism, and our firm belief is that sexual exploitation of children is a fundamental abuse of a child's human rights and dignity. For this reason, TUI Nordic has decided to contribute to the task of tackling the problem and thus protecting children from sexual exploitation. A way for the tourist sector to achieve this is by promoting good practices and self-regulation. In this work, we recognise the influence we have as a major player in the tourist sector. Even if we have no reason at all to believe that this kind of abuse has been carried out in your hotel, hotels often are the scene of the abuse . . . We believe that, by staying vigilant and taking a few simple steps, you will ensure that neither your hotel – nor its good reputation – are compromised. We would also like to take the opportunity to inform you about our position if we learn that commercial child abuse has taken place at a hotel contracted by TUI Nordic. If the abuse has been the result of an employee at the hotel acting as intermediary, we will, with immediate effect, cancel any agreement or contract.

Source: Tour Operators' Initiative for Sustainable Tourism Development: Good Practices Case Studies. www.toinitiative.org/good_practices/case_studies.htm.

'however, this decrease comes at the expense of other countries, particularly in Central America, where ECPAT is only now getting more strongly involved'.[65] This might be seen as evidence of the existence of the 'tidal wave of sexual exploitation periphery' discussed earlier in this section. It is also noteworthy that, despite its fine intentions with regard to the human rights and dignity of children, in keeping with the corporate lobby in general, the company promotes self-regulation – see Chapter 2. If there is any field of tourism activity in which it is patently clear that statutory legislation rather than voluntary self-regulation is required, it is that of sex tourism.

Concluding remarks

In this chapter we have examined the aspect of irresponsibility in the tourism industry that is associated with the exercise of power through sex. Established attitudes, *machismo*, are an important influence on sexual practices, but are themselves subject to change through factors like the creation and/or removal of work opportunities arising from the effects of the current round of economic globalisation. The relationships of power between women and men, however, are not as simple as common expectations would have us believe. The LGBT community, for instance, attempts to break down the barriers which prevent its general acceptance, but faces an opposition entrenched in *machismo*. We have left unanswered the question of whether the established roles of women and men are changing as a result of these factors; but we suggest that it is worth watching this space in future decades.

There is no doubt, however, that the still currently prevailing attitudes towards and expectations of gender roles in the LAC countries lead to widespread gender-based discrimination. That such discrimination is still alive and well today is confirmed by the United Nations Population Fund (UNFPA): 'gender-based discrimination and violence pervade almost every aspect of life, undermining women's opportunities and denying them the ability to fully exercise their basic human rights'.[66] Coupled with the hardship and changing job market that is the product of economic globalisation, these factors lead to increasing migration, especially of males, in search of economic opportunities, the splintering of family units, a greater burden on women as heads of households, greater poverty and greater pressure on family members to earn illicit income through prostitution.

These trends have been accompanied by increasing international and national tourism, and, whilst the two things are not necessarily causally related, a clear association between sexual exploitation and tourism has been demonstrated in many places. But complications in understanding the phenomenon of sex tourism arise with the incidence of females indulging in sex tourism in search of male gigolos. Additionally, without minimising the effects of sexual exploitation on women, one of the most degrading aspects of the sex tourism industry is its exploitation of children. In this regard, the work of organisations like Casa Alianza and ECPAT warrant further support and finances – although not the limited, biased and conditional support recently proffered to other organisations by President George W. Bush – and the activity itself demands condemnation by all. Awareness of its dangers and appalling effects is crucial to anyone who wishes to travel responsibly.

Although only a small proportion of the whole tourism industry, sex tourism is nevertheless significant, not simply to those who earn the money or those who benefit from the physical relief, but also financially to the major commercial players in the industry. The latter can play an important role in combating sexual exploitation in the tourism industry.

Finally, we examined the health costs behind the practice of casual and paid-for sex. In contemporary society, few can be unaware of these. STDs are a major hazard associated with the activity, and HIV/AIDS has become, in common perception, a major scourge of modern society. We have introduced, but not passed judgement on, the debates surrounding the diagnosis or mis-diagnosis of AIDS and have offered references and sources where the reader can take the debate further.

Notes

1   Bruce Harris (1999) Presentation to the 24th Session of the United Nations Working Group on Contemporary Forms of Slavery. Geneva, Switzerland, June.

2   Patrick Welsh (2001) *Men Aren't from Mars: Unlearning Machismo in Nicaragua.* London: Catholic Institute for International Relations (CIIR), p. 15.

3   Montenegro, S. (2000) 'Cultura Sexual en Nicaragua', quoted in *ibid.*, p. 16.

4   *Ibid.*, pp. 16–17.

5   *Ibid.*, p. 14.

6   Vanessa Baird (2001) *The No-Nonsense Guide to Sexual Diversity* Oxford: New Internationalist Publications, p. 81.

7   *Ibid.*, p. 81.

8   *Ibid.*, p. 81.

9   *Ibid.*, p. 79.

10 Polly Pattullo (2005) *Last Resorts: The Cost of Tourism in the Caribbean*, Second edition, London: Latin America Bureau, p. 109.

11 Edward Said (1991) *Orientalism*, London: Penguin, p. 25.

12 Sanchez Taylor, J. (1999) 'Embodied Commodities', *Tourism in Focus* 30, Winter 1998/9. London: Tourism Concern.

13 Franck Michel (2006) 'Mainstreaming Holiday Sex and the Neo-colonial Attitude', in the TIM-Team Clearinghouse, 28 August.

14 Reported by Henry Chu (2005) 'Brazil: Cedí Kina of Tourism Boom', *Los Angeles Times*, 9 February.

15 M. Shamsur Rabb Khan (2006) 'Protecting Children from the Onslaught of Sex Tourism', TIM-Team Clearinghouse, 20 December.

16 Save The Children Fund (1993) *Consultation Seminar on the Sexual Exploitation of Children and the Connection with Tourism and International Travel.* Report of Proceedings. Westminster, November 1993.

17 Commission for the Defence of Human Rights in Central America (CODEHUCA) (1999) *Report of the Human Rights Situation in Costa Rica.*

18 Richards, M. (2003) 'Costa Rica: Child Exploitation', *Mesoamerica* 22, 7: 8. July. San José: Institute for Central American Studies.

19 Barger Hannum, A. (2003) 'Sex Tourism in Latin America', *Revista: Harvard Review of Latin America: Tourism in the Americas*, Spring.

20 Martha Honey (1999) *Ecotourism and Sustainable Development: Who Owns Paradise?*, Washington DC: Island Press, p. 213.

21 Bruce Harris (1997) 'Casa Alianza Warns that Central America Is New Sex Tourism Destination', www.catwinternational.org/fb/CAmerica.html.

22 Szeman, I. (2001) 'Remote Sensing: An Interview with Ursula Biemann', www. humanities.mcmaster.ca/~szeman/biemann.htm.

23 *Ibid.*

24 Cited in *New Frontiers*, the journal of the Tourism Investigation and Monitoring Team (Bangkok) in an article entitled 'The Scourge of Child Sex Tourism', September/ October 2003.

25 Polly Pattullo (2004), p. 113.

26 Harris, B. (1999).

27 Franck Michel (2006).

28 Vanessa Baird (2001).

29 For instance, Casa Alianza was particularly prominent in urging the tourism business sector to commit itself against the commercial sexual exploitation of children at the Regional Consultation on Sexual Tourism held in Costa Rica in May 2003.

30 From a Casa Alianza Rapid Response posting entitled 'Child Pornography Threatens Central America's Children', 4 February 2004, rapid-response@casa-alianza.org.

31 Franck Michel (2006).

32 Malone Banjui, A. (1994) 'Love Conquers All on Gigolo Coast Where Sex with Toy Boys Is Banned', London: *The Guardian*, August. O'Connell Davidson, J. (1998) *Prostitution, Power and Freedom*, Cambridge: Polity Press. Sanchez Taylor, J. (1999)

'Embodied Commodities', *Tourism in Focus* 30, Winter 1998/9. London: Tourism Concern. Sey, A. (2001) 'It's a Good Job – All I Have to Do Is Befriend Women Tourists', *Being There*, Summer. London: Tourism Concern. Julie Bindel (2003) 'The Price of a Holiday Fling', London. *The Guardian Weekend*, 5 July.

33 Julie Bindel (2006) 'This Is Not Romance', London *The Guardian*, 8 August.
34 Julie Bindel (2003), p. 19.
35 *Ibid.*, p. 14.
36 Sey, A. (2001), p. 34.
37 Julie Bindel (2003), p. 18.
38 *Ibid.*, p. 19.
39 Sey, A. (2001), p. 34.
40 Polly Pattullo (2005), pp. 110–11.
41 Whilst the relationships so described are the norm, it also needs to be pointed out that a less frequent but not insignificant feature of the industry is the way in which either male or female prostitutes occasionally form a permanent relationship with one of their clients leading to marriage or permanent partnership, and leading in turn to migration to the home of the client. From such a position, remittances can be and often are sent back to the family of the former service provider in what amounts to a subtle twist in the balance of power.
42 O'Connell Davidson, J. (1999) 'Doing the Hustle', *Tourism in Focus*, Winter 1998/9. London: Tourism Concern.
43 UNFPA website (2004) www.unfpa.org.
44 www.avert.org.
45 Stockwell, J. (undated) 'The Case that SEX and HIV Absolutely Do Not Cause AIDS!', Wecan Publishers.
46 www.duesberg.com.
47 *Washington Post*, 30 April 2000.
48 Prior to October 2006, South African President Thabo Mbeki had questioned whether AIDS was caused by HIV and expressed doubt and scepticism about the safety and effectiveness of ARV drugs.
49 www.mouthshut.com. www.truthcampaign.ukf.net/articles/health/cureforalldiseases. html (for article by Ivan Fraser). www.virusmyth.net/aids/data/rhaids.htm (for article by Dr Robert Herron).
50 Marcelo Ballvé (2003) 'US or Brazil to Lead World's Fight against HIV?', *NACLA Report on the Americas*, XXXVII, 2, Sept./Oct. New York. (Originally from *Pacific News Service*.)
51 *Ibid.*
52 Polly Pattullo (2005), p. 114.
53 Prosser, R. (1994) 'Societal Change and the Growth in Alternative Tourism', in E. Cater and G. Lowman (eds) *Ecotourism: A Sustainable Option?*, Chichester: Wiley, pp. 24–5.
54 Luis Cuadra, quoted in Zambrana, E. (2003) 'Prostitución infantil, entre la ruina y la deshumanización', La Prensa, Managua, 16 January: 6B–7B.
55 Romero, E. (2003) 'Hoteles disfrazados para prostituir a menores', *La Prensa*, 16 January, Managua, p. 7B.
56 www.avert.org/aidslatinamerica.htm.
57 *Ibid.*
58 WTO/OMT (1995) WTO Statement on the Prevention of Organised Sex Tourism. 11th General Assemby of the WTO/OMT at Cairo, Egypt, October.
59 Julian Borger (2005) 'Moral Ties Attached to US AIDS Cash', London *The Guardian*, 9 July.
60 Jacobson, J. (2003) 'Bush Brings Double Standard to UN, Groups Charge', 23 September, www.oneworld.net/article/view/68722/1/.
61 Nick Wadhams (2003) 'Bush Stand on Sex Trade Meets Skepticism', London: *The Guardian*, 25 September.

62 Jeevan Vasagar and Julian Borger (2005) 'Bush Accused of Aids Damage to Africa', London: *The Guardian*, 31 August.
63 Stephen Bates (2005) 'Catholic Aid Agency Faces Cash Boycott', London: *The Guardian*, 5 February.
64 Pablo Soto (2006) 'The Right to Life', *Interact*, Autumn. London: Progressio.
65 Tour Operators' Initiative for Sustainable Tourism Development: *Good Practices Case Studies*. www.toinitiative.org/good_practices/case_studies.htm.
66 UNFPA website (2004) www.unfpa.org.

8 Power and responsibility in tourism: know your place

At this endpoint in the book, the easiest concluding option to follow would be to give a round-up of the major attractions of the subcontinent, a little summary of each chapter's findings and wish everyone '*buen viaje*' in their travels. We could conclude, as does an editorial piece in *Roughnews*, the magazine for users of the *Rough Guides*, that:

> the continent is attracting increasing numbers of travellers; for despite relatively high air fares, most of the continent is cheap, its politics largely stable and more or less democratic, landscapes spectacular and cultures relatively accessible, whatever your budget. So far, too, South America has largely avoided being sucked into the vortex of the so-called war on terrorism. Long may that continue.[1]

Whilst a summary such as this may have its uses, some of its assumptions are arguable. Its most important omission, from our point of view, is the fact that it says nothing of the human relationships at the basis of virtually all tourism experiences, even those which have such a strong emphasis on the observation of physical nature rather than human nature. In other words, it says nothing of the relationships of power behind all tourism interchanges.

Throughout this book we have stressed the importance of power in the activity of tourism – the structure of power, relationships of power and the exercise of power. Power is important at all levels and all scales, from the individual tourist's decisions about the use of their money, the tour operator's choice of guides and hotels, the national government's investment policies, to the World Tourism Organisation's promotion of global tourism infrastructure – all of these and many more are examples of the ways in which power manifests itself through tourism. As Cheryl Shanks remarks, 'Tourism pretends to be apolitical, but it encapsulates problems of power and worth on a grand and global scale'.[2]

The fact that power structures and relationships lie at the heart of tourism and its development or promotion is no great revelation. They are at the heart of all economic activities. Acceptance of this leads us to the final and crucial question for this book: how can power be exercised responsibly? If power is exercised in such a way that the benefits and advantages it brings for some are counterbalanced by disbenefits and disadvantages for others, then the responsibility of the tourism activity is compromised; and the enjoyment and pleasure of tourism and travelling come at the expense of others.

So, how are we expected to exercise power in a responsible manner? Can we do so in a way that is not patronising, condescending, moralising, prescriptive, boring, repetitive, violent and repressive? As Jost Krippendorf did over twenty years ago,[3] we have tried to give positive examples and case studies as well as negative ones which lead us to question our participation in tourism. But it has not been easy for there are precious few activities of any kind which bring benefits to all those involved without incurring disbenefits of some kind. The WIN–WIN, WIN–LOSE, LOSE–LOSE dichotomies are far too simplistic, stemming as they do from the narrow, short-sighted and jargon-filled perspective of the world of business consultancies. Perhaps too many of our examples have illustrated irresponsibility in action. But we have chosen them in order to raise questions about issues of responsibility in tourism.

To many of these questions there are no cut-and-dried answers. We ask simply that the tourist, guide, hotelier, tour operator, government planner, investor and politician should be aware of the fact that they are exercising power when they decide to visit, to build, to patronise, to service, to supply, to plan, or to invest in a particular resort or activity. Moreover, as far as it is possible, we consider it appropriate, if not necessary, to be aware of the power exercised over tourism activities by the international financial institutions such as the World Bank, IMF and the supranational agencies such as the WTO/OMT, WTO/OMC, and United Nations agencies.

These are agencies and institutions which are seemingly beyond the control of the individual tourist, service provider and even of small-scale tour operators. But decisions made at this remote level are decisions and issues over which some individuals, even if only a few, have control. It is too common a belief that because these individuals operate at a global level they are therefore aware of all the issues behind and ramifications of their decisions. Not so. They may be well-informed of ideas, developments and problems at the level at which they work; but they are often ignorant of the impacts of their decisions at the local level. There are no better examples of this than those provided by the World Bank and IMF, which have a long history of local community, regional and national protests against their programmes and projects. Recently these international financial institutions, and others such as the European Bank for Reconstruction and Development, have moved into the promotion of tourism, especially through the development of the infrastructure deemed necessary for tourism in previously remote and inaccessible areas. The World Social Forum tourism egroup describes their approach to capital investment in such areas as something akin to the slash-and-burn approach to agriculture in that they invest capital to build roads which allow the invasion of tourists. By the time the problems of environmental degradation, social divisions and over-use of resources become apparent, the capital investor has moved on to another remote area.[4]

One relatively recent example of this difference between the promotion of supposedly responsible tourism at the global level and protests against its local effects was provided by the United Nations' 2002 International Year of Ecotourism (IYE). In the run-up to the IYE, the WTO/OMT along with the United Nations Environment Programme (UNEP), both bodies charged with organisation of the IYE, were pitted against a loosely organised and co-ordinated group of NGOs which dealt with issues of tourism involving indigenous groups that objected to the event's overwhelming commercial bias and insisted that it should be renamed the International Year of Reviewing Ecotourism (IYRE – an even less pronounceable acronym). The IYE made efforts to take on board some of the issues raised by the IYRE, but it remained the IYE. Of course, the issue here was not about acronyms and initials. The concerns of the group of NGOs revolved around the ways in which the tourism industry exploits indigenous groups for the pleasure and entertainment of the tourist and the profit of the tour companies to the loss of dignity and the zooification of the indigenous groups concerned – see Chapter 5.

The individual charged with responsibility of the IYE was Oliver Hillel, UNEP's co-ordinator for the IYE, who, despite his polite and detailed responses to the group of NGOs, took the decision to keep the IYE as the IYE and to retain the perspective of the tourism business community as the predominant one. At the very least it can be said that he was informed about the protestors' views as a result of the lobbying that they undertook. If this made itself felt in even a tiny way through changes in the programme of the IYE, then it serves as an object lesson to us all that individual lobbying of other individuals (especially at the level of the global decision-makers) is worthwhile.

This might sound a little like clutching at straws, but those who exercise power at a global level are not immune to criticism or lobbying. Nor are they as well informed about the effects of their decisions as we would like to believe. However powerless those at the disadvantaged side of a tourism development or project may feel, it is still an important first step to inform the decision-makers of these disadvantages. Even at its grandest scale the world is made up of individuals, and whilst the most powerful of them may seem inaccessible at times, not taking that first step renders later and more militant steps and actions less justifiable.

As well as recognising the power that we can wield as individuals, it is important to acknowledge that power differentials operate and are significant at all levels. Chapters 2 and 3, for instance, illustrate the differences in the exercise of power at, respectively, the global level and the local level. There are of course many levels in between these two and we would have wished to illustrate them all. For instance, although we have made references throughout to the work of the NGO sector in the field of tourism development, we should have liked to dedicate a chapter to their particular ways of both exercising power and combating the power of others. Likewise, the role of national governments deserves a chapter of its own even though we have made mention of their role in many of our examples.

The differentials of power, however, are also felt within each level, and it is part of the awareness of power to realise that at the community level, for instance, some community members may benefit from tourism activities while others may feel excluded. Despite awareness of these differences and the possibility of exclusion for some, it is difficult not to agree with Anita Roddick's encouragement to us all to 'Support the local community by visiting markets and listening to local music in the local cafés and bars, and avoid buying anything in the resort hotels. There's a magic to discover in the local.'[5] This is a sentiment that fits well with the 'localisation' movement,[6] an approach that seeks to counterbalance or even reverse the trends associated with globalisation. But the exhortation to avoid the resort hotels ignores the likelihood that local people are also employed in such places.

For all these different levels of power we hope that in each chapter's notes we have provided sufficient references to the work of others for the reader to pursue their interest further if they wish.

Our awareness of the exercise of power at all levels is matched by the need to inform ourselves, as those who are exercising power, of the context and background of the destination end of our tourism activity or development. This approach should help us to make the visit, the experience, the tour, a two-way exchange rather than simply one-way and for one purpose. Despite financial leakage, the economic return to the destination end is always there, but a two-way exchange implies something more than just swapping money for fun. It implies an exchange of dialogue, understanding, and an empathy with each other, regardless of our differing opinions, circumstances, values and perspectives.

As well as making for a genuine exchange, an informed position will also lead us to recognise that the tourism industry has both harmful and beneficial impacts. In turn, this will lead us, deliberately, to try to enhance the beneficial effects and to reduce the harmful effects of our visit or tour.

Information, beyond the superficial, about the place we are to visit or the development we are to promote will inevitably and by definition reduce the strange ability we all seem to have of making snap judgements about foreign places and people. Few tourists, however, manage to 'do their homework' and inform themselves about their holiday destination. Perhaps we should not be too surprised by this – the holiday is after all meant to be a change from our usual routine; we want to relax; and we do not normally wish to be burdened with the chore of reading and researching as if we were to sit an examination. (Of course, more than a few people would not perceive such discovery and learning as a chore.) But let us take it as understood that the majority of tourists are fairly ignorant of life beyond the tourism experience in their chosen destination, and even those who do take the trouble to learn before their visit are relatively ignorant of many of the nuances of life for the resident population. It is this ignorance which leads us to make judgements according to the reference point of our own home environment. For all those First World tourists who manage to visit the LAC countries, using the home environment and culture as a basis for one's judgements and decisions is clearly false and will inevitably lead to mistakes. Yet we all do it. It is so easy to fall into the trap because we are in a strange environment, one where at times we feel pressured into making snap judgements and decisions, a proportion of which turn out to be wrong or embarrassing and of which we later become ashamed.

It is easy to assume that such poor judgements and decisions are made only by inexperienced and relatively ignorant young travellers, especially gap year students for whom the trip is an essential experience-builder. But strangeness in the LAC countries is just as likely to be felt by wealthy nature travellers and culture vultures from the North who have the money to surround themselves with First World expectations, trappings and conditions in order to reduce the probability of their committing the equivalent of social gaffes or bad decisions.

Our gentle mocking of gap year students – or 'gappers', as they are sometimes called – may be misplaced. Many of them are fully aware that they are travelling for their own experience and that they – not the locals – will be the beneficiaries of their travels. There are plenty of exceptions to this of course, but many gap year travellers travel deliberately to get close to the local people and have no wish to exercise power over them. Indeed, many pretend and claim to be poor, thereby expressing an alliance and empathy with many of the populations they visit. Increasingly, many gap year students seek to work or volunteer on development projects in order to contribute something positive to the places they visit. Without ever having travelled before (except within the confines of the family), they appear to be aware of the difference between a tourist visit and an expression of solidarity; and, despite the occasional bouts of brash hedonism and selfishness, many young people exercise a degree of responsibility, the need for which older generations were simply unaware of at the same stage of personal development. But this solidarity may not be strong enough as a manifestation of responsibility. Certainly Hutnyk is seeking more when he suggests that:

> It is not enough just to raise questions about the moral propriety of First World youth taking holidays among the people of the Third World; it is not enough to encourage discussion of such contradictions in cafes along the banana-pancake trail. Nor is it sufficient to reflect critically upon the politics of charity while working – because something must be done – at a 'sound' street clinic.[7]

Despite the significance of the point that Hutnyk is making, we consider that, if this greater awareness in the population of young tourists (or travellers as they would prefer to be called) has arisen from a higher media profile of the problems caused by the tourism industry or of more general environmental and social problems, then we see that as an encouraging sign. This is not to excuse some of the splendid stupidity displayed by gap year students, young travellers and volunteers in the LAC countries; but far worse than this is the ignorance and prejudice of the profiteering companies dedicated to selling them their experiences.

Still on the theme of informing ourselves before we travel, it is worth making the point that this era of electronic information that we now live in has at least made the information we seek about our holiday destinations a great deal more easily and widely accessible. Of course there are no fewer problems with the information available now on the internet than there were when we had to rely on the glossy brochures provided by travel agents. Search on the internet for almost any particular resort in the LAC countries and all you will find are glossy, well-illustrated accounts singing the praises of the said resort. It takes somewhat deeper digging and detailed knowledge of appropriate websites to find anything that might be considered to be a critical analysis. In the Appendix we have tried to give as full a list of relevant websites as possible, but we have not simply given the sites of those organisations generally considered to be critical of international tourism. Also included are the website addresses of supranational organisations and agencies which collectively see their task as that of promoting the international tourism industry and of reducing all barriers to its promotion and growth, even where it is clear that such policies can only promote forms of tourism which will serve to disbenefit some of the players in the activity. To cause such disbenefits and disadvantages is irresponsible, we believe, but at least it is done by most such organisations in genuine but ill-informed belief that their policies are appropriate and responsible. It is correct therefore that we should at least be aware of these perspectives. It is even more important, however, that we should be aware of blatantly deliberate and irresponsible attempts to sell us a 'product', a destination, a tour, a hotel, an eco-lodge by means of fictitious claims – in other words, by what is commonly known as 'greenwash' (see Chapter 4). Information from the tourism industry contains a great deal of greenwash, and it is crucial that we inform ourselves about how to spot it, avoid it, report it (to the guidebook publishers, for instance) and counter it.

We did not wish to end this book by giving a list of DOs and DON'Ts, a prescription for how to 'do' responsible tourism correctly. Despite the best efforts of The International Ecotourism Society (TIES) and the (almost) manuals that they produce on how to practise and promote ecotourism, there is no clearly defined way to practise tourism responsibly – it varies according to the values of, and many other factors associated with, those who define it. We have no wish to be prescriptive in this book. It is not our part to tell nor suggest to the reader that they should or should not do certain things. But few people would object to the idea that whilst on holiday we should not only relax and have fun but should also learn a little about the place and people that we are visiting. It is not such a bad idea that we should return home a little wiser and more knowledgeable about the world than we were when we left home.

Naturally, it is also not such a bad idea that we do not lord it over those we visit. The First World status of the visitor is no indication of superiority; we should take it instead purely as an indication of power, the economic power to take such a trip. To confuse power with superiority is a mistake. In travels around the LAC countries it is a common experience to come across groups of young or old, especially male, visitors who have adopted unquestioningly a belief in the inferiority of those they visit. This is racism, and what is so surprising about it is that so many tourists do their utmost to justify it. Under the heading 'The Dark Side of Tourism', organisers of the tourism egroup at the World Social Forum in India in January 2004 suggest that many travellers are now effectively pilgrim tourists who 'travel more to enforce bias and bigotry rather than expand knowledge and mind'.[8]

If we are travelling to learn as well as to have fun and relax, then it should be self-evident that learning can take place only under humility. Arrogance and prejudice reinforce ignorance, which in turn reinforces arrogance and prejudice, attitudes which together make learning impossible.

Notes

1   Richard Trillo and Niki Hanmer (2004) Editorial in *Roughnews*, London: Rough Guides, Summer, Issue 23, p. 2.
2   Cheryl Shanks (2002) 'Nine Quandaries of Tourism', in *Revista: Tourism in the Americas – Harvard Review of Latin America*, Winter, www.people.fas.harvard.edu.
3   Jost Krippendorf (1987) *The Holiday Makers: Understanding the Impact of Leisure and Travel*, Oxford: Butterworth-Heinemann.
4   World Social Forum tourism egroup (2003) 'The Dark Side of Tourism', a report in preparation for tourism-related events for the World Social Forum in Mumbai, India, January 2004, Tourism Investigation and Monitoring Team, Bangkok.
5   Anita Roddick (2004) 'Fair Play', interviewed by Richard Hammond in London: *Guardian Travel*, 22 May 2004, p. 15.
6   Localisation is a movement associated with Colin Hines, especially through his book *Localization: A Global Manifesto*, published in 2000 by Earthscan (London).
7   John Hutnyk (1996) *The Rumour of Calcutta: Tourism, Charity and the Poverty of Representation*, London: Zed Books.
8   World Social Forum tourism egroup (2003), http://wsf-tourism.org.

Appendix
Websites related to travel and tourism

Association of Caribbean States (ACS) www.acs-aec.org Sustainable tourism site www.acs-aec.org/tourism.htm

Association of Independent Tour Operators (AITO) www.aito.co.uk/ including 'responsible tourism' www.aito.co.uk/corporate_Responsible-Tourism.asp

Best Foot Forward www.bestfootforward.com

Big Volcano Ecotourism Resource Centre www.bigvolcano.com.au/ercentre/ercpage.htm

Business Enterprises for Sustainable Tourism (BEST) www.sustainabletravel.org/

Business Charter for Sustainable Development www.iccwbo.org/policy/environment/id1309/index.html

Campaign for Environmentally Responsible Tourism (CERT) www.c-e-r-t.org

Campaign for Real Travel Agents (CARTA) www.realholiday.co.uk

Caribbean Alliance for Sustainable Tourism (CAST) www.cha-cast.com/

Certification in Sustainable Tourism Program of Costa Rican Tourism Institute www.turismo-sostenible.co.cr/EN/home.shtml

Choose Climate website www.chooseclimate.org

Conservation International (Ecotourism) www.ecotour.org/xp/ecotour/

Corporate Watch www.corporatewatch.org/

Cultural Survival www.culturalsurvival.org/

Department for International Development (DFID) www.dfid.gov.uk/

Development Planning Unit, University College London www.ucl.ac.uk/dpu/

Economic Commission for Latin America and the Caribbean (ECLAC) www.eclac.cl/ ECLAC site on Sustainable Development and Human Settlements Division www.eclac.org/dmaah/

End Child Prostitution, Pornography and Trafficking (ECPAT) www.ecpat.org/

Environmental News Network www.enn.com/

Ethical Volunteering www.ethicalvolunteering.org/

Exodus responsible tourism site www.exodus.co.uk/restourism.html

Explore! responsible tourism site www.exploreworldwide.com/worldwide/responsible tourism.jsp

Fair Trade in Tourism Network www.egroups.com/group/fairtradetourism/

Footprint guidebooks www.footprintbooks.com/

Forum for the Future www.forumforthefuture.org.uk/

Friends of the Earth (FOE) Corporate Accountability Campaign www.foe.co.uk/campaigns/corporates/

Green Globe worldwide benchmarking and certification program www.greenglobe21.com/

Greenpeace (UK) www.greenpeace.org.uk/

Guardian Tourism Green Travel http://travel.guardian.co.uk/tag/green

Institute for Central American Studies (ICAS) for Mesoamerica journal www.mesoamericaonline.net/

Inter-American Development Bank (IDB) www.iadb.org/

International Business Leader Forum (IBLF) www.iblf.org/ and Tourism Partnership www.tourismpartnership.org/

International Centre for Responsible Tourism www.icrtourism.org/

International Ecotourism Club http://ecoclub.com/

International Ecotourism Society www.ecotourism.org/

International Institute for Environment and Development (IIED): Tourism site www.iied.org/SM/tourism/index.html Human Settlements site www.iied.org/HS/index.html

International Porter Protection Group (IPPG) www.ippg.net/

International Union for Conservation of Nature and Natural Resources (IUCN) World Conservation Union www.iucn.org/

International Year of Ecotourism 2002 www.worldtourism.org/sustainable/2002 ecotourism/eng.htm

National Geographic Center for Sustainable Destinations www.nationalgeographic.com/travel/sustainable/index.html

North American Congress on Latin America (NACLA) www.nacla.org/

Overseas Development Institute (ODI) Tourism Programme www.odi.org.uk/propoortourism/index.html

Panos Institute www.panos.org.uk/

Planeta: global journal of practical ecotourism www.planeta.com/

Progressio (formerly the Catholic Institute for International Relations, CIIR) www.progressio.org.uk

Pro-poor Tourism Partnership www.propoortourism.org.uk/

Rainforest Alliance sustainable tourism www.rainforest-alliance.org/tourism.cfm?id=main

Redturs: network of communitarian tourism of Latin America www.redturs.org/

René Waksberg's Tourism Research Links www.waksberg.com/research.htm

Responsible Hospitality Institute www.hospitalityweb.org/

Responsible Tourism Partnership www.responsibletourismpartnership.org/

Responsibletravel.com www.responsibletravel.com/

Survival International www.survival-international.org/

Sustainable Tourism – Eliminating Poverty (ST-EP) WTO/OMT www.unwto.org/step/index.php

Swedish International Development Co-operation Agency (SIDA) www.sida.se/

Tearfund www.tearfund.org

Third World Network Tourism Investigation and Monitoring Team (T.I.M.-Team) www.twnside.org.sg/tour.htm

Tour Operators Initiative for Sustainable Tourism Development www.toinitiative.org/ For good practice case studies www.toinitiative.org/good_practices/case_studies.htm

Tourism Concern www.tourismconcern.org.uk/

Travel Foundation www.thetravelfoundation.org.uk/

United Nations Development Programme www.undp.org/

United Nations Environment Programme (UNEP) Tourism www.uneptie.org/pc/tourism/policy/home.htm

United Nations Educational, Scientific and Cultural Organisation (UNESCO) http://portal.
unesco.org/en/ Also: UNESCO World Heritage Centre http://whc.unesco.org/
United Nations Human Settlements Programme (UN Habitat) www.unhabitat.org/
Urban Archaeological Heritage of Latin America, International Council on Monuments
and Sites (ICOMOS) www.icomos.org/studies/latin-towns.htm
Voluntary Service Overseas (VSO) www.vso.org.uk
World Conservation Monitoring Centre (WCMC) www.wcmc.org.uk
World Development Movement (WDM) www.wdm.org.uk/
World Ecotourism Summit 2002 Québec Declaration on Ecotourism www.world-
tourism.org/sustainable/IYE/quebec/anglais/declaration.html
World Land Trust www.worldlandtrust.org/
World Tourism Organisation (WTO/OMT) www.unwto.org/ Also: WTO/OMT Global
Code of Ethics www.unwto.org/code_ethics/eng/global.htm and WTO/OMT-
Sustainable Development of Tourism www.world-tourism.org/sustainable/news.htm
World Trade Organisation (WTO/OMC) www.wto.org/
World Travel & Tourism Council (WTTC) www.wttc.org/
World Wide Fund for Nature-UK (WWF-UK) www.wwf.org.uk/ Also: WWF-UK on
tourism www.wwf.org.uk/researcher/issues/Tourism/index.asp
ZMagazine www.zmag.org

Bibliography

The Endnotes at the end of each chapter in this book are extensive and cover many references that the reader, we hope, will find useful. We realise, however, that readers may wish to refer to a short list of major works of significance rather than sifting through a mountain of, sometimes obscure, references. For this reason, we have included here a bibliography of major works that we consider to be of significance to the subject matter covered in this book. But we wish to emphasise two important points: first, if works are not listed here it does not mean that they are of no significance; second, our treatment of the subject matter has been inter-disciplinary and in a brief list such as this it is impossible to cover all the significant works. Incomplete though it is, we suggest the following for further general reading in this field.

Bauman, Z. (1998) *Globalisation: The Human Consequences*, Cambridge: Polity Press.

Butcher, J. (2003) *The Moralisation of Tourism: Sun, Sand . . . and Saving the World?* London: Routledge.

Cater, E. and Lowman, G. (eds) (1994) *Ecotourism: A Sustainable Option?* Chichester: Wiley.

Cooke, B. and Kothari, U. (eds) (2001) *Participation: The New Tyranny*, London: Zed Books.

De Soto, H. (2001) *The Mystery of Capital: Why Capitalism Triumphs in the West and Fails Everywhere Else*, London: Black Swan.

Duffy, R. (2002) *A Trip Too Far: Ecotourism, Politics and Exploitation*, London: Earthscan.

Easterley, W. (2001) *The Elusive Quest for Growth: Economists' Adventures and Misadventures in the Tropics*, Cambridge, MA: MIT Press.

Escobar, A. (1995) *Encountering Development*, Princeton: Princeton University Press.

France, L. (ed.) (1997) *The Earthscan Reader in Sustainable Tourism*, London: Earthscan.

Galeano, E. (1997) *Open Veins of Latin America: Five Centuries of the Pillage of a Continent*, New York: Monthly Review Press.

Ghimire, K. B. (ed.) (2001) *The Native Tourist: Mass Tourism Within Developing Countries*, London: Earthscan.

Honey, M. (1999) *Ecotourism and Sustainable Development: Who Owns Paradise?* Washington DC: Island Press.

Hong, E. (1985) *See the Third World While It Lasts: The Social and Environmental Impact of Tourism with Special Reference to Malaysia*, Penang: Consumers' Association of Penang.

Krippendorf, J. (1987) *The Holiday Makers: Understanding the Impact of Leisure and Travel*, Oxford: Butterworth-Heinemann.

Lea, J. (1988) *Tourism and Development in the Third World*, London: Routledge.

McLaren, D. (2003) *Rethinking Tourism and Ecotravel: The Paving of Paradise and What You Can Do to Stop It*, Second edition, Bloomfield, CT: Kumarian Press.

Mowforth, M. and Munt, I. (2003) *Tourism and Sustainability: Development and New Tourism in the Third World*, Second edition, London: Routledge.

Pattullo, P. (2005) *Last Resorts: The Cost of Tourism in the Caribbean*, Second edition, London: Latin America Bureau.

Richards, G. and Wilson, J. (eds) (2004) *The Global Nomad: Backpacker Travel in Theory and Practice*, Clevedon: Channel View Publications.

Sen, A. (1999) *Development as Freedom*, Oxford: Oxford University Press.

Sharpley, R. and Telfer, P. (eds) (2002) *Tourism and Development: Concepts and Issues*, New York: Channel View Books.

Smith, M. and Duffy, R. (2003) *The Ethics of Tourism Development*, London: Routledge.

Smith, V. (ed.) (1989) *Hosts and Guests: The Anthropology of Tourism*, Philadelphia: University of Pennsylvania Press.

Smith, V. L. and Brent, M. (eds) (2001) *Hosts and Guests Revisited: Tourism Issues of the Twenty-first Century*, New York: Cognizant Corporation.

Theroux, P. (1979) *The Old Patagonian Express: By Train Through the Americas*, London: Penguin Books.

Index

Acapulco 125
Action Aid 45
Agnew, M. 118
AIDS Relief Plan (USA) 218
Amazon Region Protected Area (ARPA) 122
Amazon Tours 151
Amnesty International 47
Andean trekking porters 157–9
Angotti, T. 172, 175
Annan, Kofi 19
anti-globalisation movement 28, 32, 33
Ashley, C. 3, 60
Asian Development Bank (ADB) 184
Association of Fishermen of Barcelos 152
Association of Independent Tour Operators (AITO) 43–4
Association of Indigenous People of Barcelos (ASIBA) 152
Association of United Communities for the Economic and Social Development of the Lower Lempa in Jiquilisco (ACUDESBAL) 64–5
Astor, M. 142, 147–8
Atkinson, D. 70
Audubon Cooperative Sanctuary System 113
AVERT 212, 214

Baird, V. 198, 207
Ballvé, M. 213
Bandarin, F. 181
Bandy, J. 29
Barkin, D. 72, 124
Barton, W.D. 153
Belize Enterprise for Sustained Technology (BEST) 79, 164, 165
Bennett, J. 149, 153, 154
Berwari, N. 179

BEST guesthouse scheme 165
Bindel, J. 208, 209
Blum, W. 12, 144
Bolivar, Simon 11
Bolt, A. 89
Borger, J. 217
Brandon, K. 77
Brazilian National Foundation for Indian Affairs (FUNAI) 151
Brechin, S. 77
British Airways (BA) 121
Britton, S. 12
Brown, Gordon 19–20
Brundtland, Gro Harlem 108
Bush, George W. 216, 218
Butcher, J. 144
Butler, R. 21

CAFOD (Catholic Agency for Overseas Development) 20, 218
Cahuinari National Nature Park (Colombia) 123
Campaign for Real Travel Agents (CARTA) 44
Campos, P. 201
Cancún (Mexico) 13, 102, 125, 128, 129, 176–7
Cancún WTO/OMC meeting (2003) 33, 176
Carazo, Rodrigo 72
carbon budgets 118–21
CarbonNeutral Company 120
CARE Bolivia 70
CARE International 70
Caribbean Alliance for Sustainable Tourism 36
Casa Alianza 203–5, 220, 221n29
Castro, Fidel 176, 207
Cattarinich, X. 60

Central Intelligence Agency (CIA) 28
Centre for Health and Gender Equity 217
Certification for Sustainable Tourism
    (CST) (Costa Rica) 113–14
Chavez, Hugo 46
Che Guevara Trail (Bolivia) 68
Christian Aid 10, 35, 47
Cienfuegos, O. 79, 81
cities, crime/insecurity in 173–4; and
    cultural heritage 180–2; demonisation
    of 174–5; as economic machines
    176–9; and *favela* tours 186–9; First
    World/developing world difference
    171; and gentrification creep 181–2;
    and global urbanisation 191–2n19;
    marginalisation of 176; poverty/slums
    in 171–2, 174, 177, 179, 180–1, 184;
    and pro-poor tourism 182–6; and
    tourism development 175–6; travellers'
    tales concerning 173; and trickle-down
    regeneration 177, 187; and urbanisation
    as anti-development 173–6;
    zooification in 181
Ciudad Juárez (Mexico) 205, 206
Cleaver, F. 76
Cleverdon, R. 45
Climate Care 120
codes of conduct 37–41
coffee tours (Colombia) 130, 131
Colantino, A. 172
Colchester, M. 139
commercial sexual exploitation of children
    (CSEC) 218
Commission for the Defence of Human
    Rights in Central America
    (CODEHUCA) 202
The Community Tourism Guide 66
community-based tourism 94–5; analysis
    of 66; concept 65–6; criteria for 66;
    criticism of 70–1; and definition of
    community 67; development of 68; and
    local involvement 66–7, 72–5; and
    powerful elites 67–8; and role of
    outsiders in 71–2; and solidarity tour
    groups 68–70; and urban setting 66
Confederation of British Industry (CBI) 35
Connell, J. 105
Conservation International 71, 76
Cooke, B. 75–6
CORE Coalition 47
Corporate Social Responsibility (CSR)
    33–4, 41–4
Costa Rica 201–4

Costa Rican Federation for Environmental
    Conservation (FECON) 90
Costa Rican Institute of Tourism (ICT) 90,
    91
Crick, B. 22
Crowther, A. 158
Cuadra, L. 214
Cuba 12, 20, 172, 176, 203, 207

Davis, M. 181
De Soto, Hernando 184–5
Department of Environment, Food and
    Rural Affairs (DEFRA) 121
Department for International Development
    (DFID) 3, 58, 59, 70
Desierto de los Leones (Mexico) 122
Distintivo Ecoturistico 113
Dodds, K. 6
Doha meetings 20, 31
Dragoman 148
Duesberg, P. 212
Duncan, 174
Dwek, D. 66–7, 186–9

Eade, D. 175
Eco-Tours SA 156–7
ecolabelling/certification 113–15
ecological footprint 117
Economic Commission for Latin America
    and the Caribbean (ECLAC) 54
ecotourism 13–14, 143; as big business
    110; as contested issue 110; criteria for
    111; definition of 110–11; development
    of 111–12; lite 112–13; rapid
    proliferation of 112; and role of
    outsiders 71; scope/scale of 111
Ecotur 113
Ecuador 19, 71, 126
El Nuevo Diario newspaper 89
Elliott, L. 28
Ellwood, W. 22
emissions trading 121–2
Encounter Overland 148
End Child Prostitution, Pornography and
    Trafficking (ECPAT) 38, 203, 208,
    218–19, 220
environment 7; and air transport 107; and
    destruction of agricultural land 106;
    direct, indirect, induced impacts on
    107–8; and disturbance of habitat/
    wildlife behaviour 105; and energy
    consumption 106–7; foreign versus
    domestic tourism 130–1; and global

warming 118; and habitat loss/
landscape change 106; impacts on
101–2; nature conservation 122–5;
pressures of tourism 131–2; protected
areas 122–5; and sanitation/pollution
106; and surface transport 107; and
tourism 105–8; and visitor pressure,
erosion, wear and tear 106; and
water demand 106; and wildlife in
captivity 105–6, *see also* natural
environment
Equations 32
Escobar, A. 2
Esteva, G. 2
The Ethical Travel Guide 66, 74
European Bank for Reconstruction and
Development (EBRD) 225
European Union (EU) 5, 21, 22, 31, 121
Exodus 40–1, 42–3; Responsible Tourism
Policy 148
Explore Worldwide 40–1, 158

Fair, J. 71, 72
fair trade 44–6
Fair Trade in Tourism 41, 43
favela tourism 14, 66–7, 186–9
Finca Sonador (Costa Rica) 72–4, 82
Finnair Travel Services 39
First World 8n9, 208, 229; and benefits of
tourism 53; cities of 171; and
distinction with Third World 14; and
economic exploitation 15; and
ecotourism 13; and free trade 22; and
process of acculturation 145; and Third
World dependency 46–7; and
TNCs/national companies 27; wealth
in 5
Foucault, M. 76
Fourth World 140, 145
Fraser, B. 125
free trade 177, 199; and comparative
advantage 21; concern/resistance to 33;
and debt crisis 25; and deregulation 22;
as essential 21; ideology of 21; and
improved infrastructures 22, 24–5; and
industry constraints 35; and inequalities
of development 22; and level playing
field 22, 27, 28; local effects of 27–9;
requirements of 21–2; strategies 25, 27
Friends of the Earth 47
Fukuyama, F. 2
FUNAI (Brazilian Indian Foundation) 141
Fundación Maquipucuna 71

G7 31
G8 20, 21, 22, 34, 53, 85, 117
G8 Summit (Genoa, 2001) 33
Galápagos Islands 125, 126
Galeano, E. 11
Garifuna community (Honduras) 92–4,
137, 142–3
Gelber, George 20
General Agreement on Trade in Services
(GATS) 14, 29, 31–3
Ghimire, K. 15
Global Code of Ethics for Tourism 40
Global Environmental Facility (GEF) 122
The Good Alternative Travel Guide 66
Graham, Y. 16
Gray, L. 206
Green Globe 36, 43–4, 115
Green Globe 21 36
Green Globe Accreditation (GGA) 36
Green Leaf 115
greenwashing 14, 34, 43, 112–13
Griffiths, T. 54
Guarani-Kaiowá 142
The Guardian 217
Guatemala 12, 204
Guerrero, R. 174
Guevara, Ernesto Ché 13, 68, 70
Gunson, P. 79

Hadley Centre 118
Haiti 213
Hall, S. 137
Hamilton, A. 149, 154
Harris, B. 204–5, 207
Haslam, P.A. 36
Havana (Cuba) 172, 176
Henderson, D. 34
Henkel, H. 75, 76
Hershberg, E. 17
Hettne, B. 16
Hillel, O. 226
HIV/AIDS 195, 204, 211–18, 220
holiday footprint 117–18
Honduran Chamber of Tourism 91
Honey, M. 81, 90, 91, 110–11, 112, 203
human development index (HDI) 56
human poverty index (HPI) 56
Hunter, C. 117
Hutnyk, J. 46, 228
Hutton, W. 157

In Focus magazine 65
Inca Trail 158

Indigenous Missionary Council 141
Indigenous Organisation of the Negro River (FOIRN) 150
indigenous peoples, and assimilation 138, 141–2; attraction of 137; autonomy of 140; case studies 149–66; and cultural difference 138; and exploitation of natural resources 140–1; extermination of 138–9, 140; external influences on 142; involvement in tourism 137, 139–40, 142–4, 146–7; as key selling point 137; and local culture/tourism conflict 166; and nature conservation 125; and photography 137, 143, 147–9, 160–3; problems affecting 166–7; and transculturation 145–6; zooification process 144–5
Inka Porter Project 158
Institute for Central American Studies (ICAS) (Costa Rica) 202
Institute for the Environment and Natural Renewable Resources (IBAMA) (Brazil) 152
Institute of Environmental Protection of Amazonas (IPAAM) 152
Inter-American Development Bank (IDB) 24, 43, 92, 115, 152, 165, 180, 181–2
Inter-American Foundation 43
Intergovernmental Panel on Climate Change (IPCC) 107, 118
International Business Leader's Forum (IBLF) 41, 44
International Financial Facility (IFF) 19, 20
International Monetary Fund (IMF) 7, 10, 21, 32, 53, 57, 115, 141
international non-governmental organisations (INGOs) 70, 71, 75, 76, 95, 115, 139, 174, 200
International Porter Protection Group (IPPG) 158
International Year of Ecotourism (IYE) (2002) 14, 226
IUCN 108
Iwand, W.M. 38

Jacobson, J. 217
Jagger, Mick 144
Japan Travel Bureau 39
Johnston, B. 77
Journey Latin America (JLA) 40, 120

Kalisch, A. 31–2, 45

Kayabi, M. 142
Kellogg Foundation 43
Khan, S.R. 201
Khor, M. 25
Klein, N. 34, 37, 38
Korten, D. 28
Kothari, U. 75–6
Krippendorf, J. 225
Kuna General Congress (KGC) 153, 154
Kuna Indians (Panama) 67, 139, 144, 149, 152–5
Kyoto 121

La Higuera (Bolivia) 68, 70
Lacerda, Marcio 141
Landry, C. 180
Las Terrazas community (Cuba) 79, 81–2
Latin America and the Caribbean (LAC) 1, 4; and access to business infrastructure 22, 24–5; alignment with other countries 5, 8n9; and the conquistadors 138; defined as under-developed 21; and deregulation/privatisation 27–8; development of tourism in 10–15; and domestic tourism 130–1; and economic globalisation 4; environmental/social protections in 32–3; flows of visitors to 4; GDP 54, 95n1, 118; and growth of professional class in 4; and land disputes, evictions, displacements 63–5; and middle class culture 6, 8–9n11; as playground of the wealthy 10; poverty/inequality in 19, 54–60, 62–3; power/privilege in 10; and pro-poor tourism 58–60, 62–3; and relationship with TNCs 27–8; and tourism/poverty reduction link 53–4; and tourists in/locals out 5–7; urbanization in 171, 172, 173–4
least developed countries (LDCs) 31
lesbian, gay, bisexual and transgender (LGBT) 196
Ljung, P. 180
local participation, and community-based tourism 75; concept of 76; criticism of 75–6; as essential to sustainability, responsibility, development 75; as good 76; as means of interaction/development 77; problems with 77; scale of 82; and structures of power/politics 76–7; as successful 76; typology of 77–9, 81–2, 145, 146

Local System of Integrated Health Care (SILAIS) (Nicaragua) 214
Lonely Planet Guides 129
Longo Maï movement (Costa Rica) 72
Los Micos Beach and Gulf Resort (Honduras) 91–4

MacEwan, A. 22
McKercher, B. 35
McLaren, D. 28
McLintock, Maria 160–3
McNeil, J. 90
'Make Poverty History' (2005) 33
Managua 12
Mann, M. 66, 67, 68
Manu Park (Peru) 125
Mapuche Indians 118, 139–40, 159–60
Mar del Plata (Argentina) 12, 125, 130
Marina Puesta Del Sol (Nicaragua) 86–9
Marshall, J. 90
Marshall, O. 57
Marsteller, P. 151
mass tourism, difficulties of containing 128–30; as main problem 125, 127–8
Matthews, B. 119
Mayas 137
mega-projects 95; case studies 82–94
Menchú, R. 144
Mercosur 178–9
Mesoamerica magazine 202
Mexico City 66
Michel, F. 200–1, 207
Michoacan (Mexico) 72
Miliband, D. 120
Millennium Development Goals 3, 10, 17–21, 58
Monarch butterfly 124–5
Montego Bay (Jamaica) 13
Moody, T. 153
Morales, Evo 46
Moseley-Williams, R. 22, 58
Moser, C. 174
Mosse, D. 76
mountain biking (Mexico) 132
Mumtaz, B. 172
Mundo Maya (Maya World) project 79, 80
Muntarbhorn, V. 201
Mutal, S. 181
Mutter, M. 179

Nahuel Huapi National Park (Argentina) 111–12

National Ecotourism Accreditation Programme 115
natural environment 101; authenticity of 102; characteristics of 102; and climate 103; and coastal landscapes 103; and generalised landscapes 103; and high-profile/dramatic physical features 103; impacts on 102; and isolation of attractions 104–5; marginal significance of 105; and wildlife 104, *see also* environment
Naturally Costa Rica magazine 202
neoliberalism 21–9, 177
New Economics Foundation 41
New Internationalist magazine 54, 120
New Moral Tourism 144
non-governmental organisations (NGOs) 20, 32, 41, 43, 47, 71, 72, 95, 175, 200, 203, 226
North American Free Trade Agreement (NAFTA) 22
Nueva Esperanza (El Salvador) 68, 69, 82

Ocampo, J.A. 57
O'Connell Davidson, J 209, 210
Organisation for Economic Co-operation and Development (OECD) 21
Oriente region (Ecuador) 141
Orizzonti 39
Otishi National Park (Peru) 125
'Our Common Future' (Brundtland Commission Report, 1987) 108
Overseas Development Institute 3
Oxfam 10, 45
Oyola-Yemaiel, A. 111

Page, S. 105
Palacios, Julio 20
Papagayo (Costa Rica) 89–91
participation *see* local participation
participatory rural appraisal (PRA) 76
Pattullo, P. 82, 85, 199, 206–7, 209, 213
Pendleton, A. 35
Peres, S. 152
Phillips, T. 189
Plan Puebla-Panamá (PPP) 22, 23–4
Pleumarom, A. 70
Potter, R. 172
Prague Ministerial Conference (2000) 33
Pratt, M. 145
Premier Tours 39
Pretty, J. 77–9, 81–2, 95, 145, 146

pro-poor tourism 58–60, 62–3, 74–5, 94–5, 182–6, 191
Pro-Poor Tourism Partnership 60
Progressio 138–9, 218
'Promoting Responsible Travel in Curaçao and Bonaire' guidelines 38, 39
Prosser, R. 213–14
Punta del Este (Uruguay) 12

racism 229
Raichman, A. 213
Rainforest Alliance 115
Red Argentina de Turismo Rural (Ratur) 130
Resende, S. 15
Responsible Tourism Guidelines (2001) 43–4
Rio de Janeiro (Brazil) 186–90
Rio Negro (Brazil) 150–2
Rist, G. 16–17
Riviera Maya (Mexico) 128–9
Rocinho 186–9
Roddick, A. 227
Rojas, E. 180, 182
Rosen, F. 22
Rostow, W. 21
Ryan, C. 105

Sachs, J. 177–8
Sachs, W. 16
Said, E. 200
St Lucia Heritage Tourism Programme (HTP) 62
Salam, Makario 165
Sandals Resorts (the Caribbean) 36, 84–6
Scandinavian Leisure Group 39
Schumacher, E. 176
Seattle Ministerial Conference (1999) 33
Serra do Pardo National Park (Brazil) 123
Serrano, J.B. 202
sex, and Catholic Church 197, 218; and gender/identity 196, 198–200, 220; and machismo 196–8, 199, 219; and politics/patriarchy link 207; and power relations 195–6, 219–20; and prostitution/migration link 205; relationship with tourism 195–6
sex tourism 7, 38; advertisements for 201–2; and beachboy phenomenon 209–10; and child abuse 201, 202, 203–6, 208; economic factors 209–10; and female exploitation 201, 202, 206–7; geographical locations 201; in

LAC 201–8; numbers of participants 202–3; and Orientalism/Other 200–1; as reflection of power 200–8, 209–10; and sex education 210–11; and sexually transmitted diseases 210–19; and western cultural life 200; and women tourists 208–10
Shanks, C. 3, 224
Shaw, J. 117
Shiva, V. 55
Sierra del Rosario (Cuba) 79
Simons, J. 176
Smithsonian Institute 93
Socioenvironmental Institute (ISA) (Brazil) 141, 150
solidarity tourism 68–70
Soto, P. 218
South American Experience 41, 120
South American Handbook 12, 14
Standish, A. 144
Stewart, B. 85
Stirrat, R. 75, 76
Suresh, K.T. 32
Survival International 139, 142, 144, 166
sustainable development 1–2, 7, 33–4
sustainable tourism 108–10
Sustainable Tourism Stewardship Council (STSC) 115
Swain, M.B. 153

Tannerfeldt, G. 180, 182
Taylor, H. 77
Taylor, S. 200, 209
Tearfund 41, 43, 148
Terra do Meio Ecological Station (Brazil) 123
The International Ecotourism Society (TIES) 71, 76, 110, 229
Theroux, P. 171, 173, 179, 186
Third World 8n9, 13–15, 208; and asymmetry of power 145; as attractor/detractor for tourism 57; as botanical garden for First World 15; and contact perspective 145; debt crisis 33, 46–7; and ecological footprints 117; effect of loans/interest rates on 25; and exploitation of natural resources 140–1; and having a voice 146; impact of urban tourism on 183–4; and inequalities of development, economy, power 22; and liberalisation of trade 28; and pro-poor tourism initiatives 53; and process of acculturation 145; and

production of capital 185; and
responsible tourism 47; and structural
adjustment programmes/poverty
reduction strategies 17, 26, 32, 177;
and transculturation 145–6
Toledo Ecotourism Association (TEA)
(Belize) 72, 139, 163–6
Toledo Peoples' Eco-Park Development
165
Torres, R. 128
Tour Operators' Initiative (TOI) for
Sustainable Tourism Development 38,
39, 40
'Tourism: Unfair Practices – Equitable
Options' conference (2003) 38
tourism, alternative 172; and authenticity
144; best practice initiatives 38–40; as
binge-pleasuring 63; challenges/
criticisms of 59–60; contribution to
local economies 27; domestic/foreign
difference 6, 9n12; as form of
migration 5–7; and fragmentation
of industry 35–6; and indigenous
groups 1; and information 227–9;
issues-based 1; judgements/decisions
concerning 227–8; local
problems/effects 63, 65; models 21;
and nature tourists 1; and notion of
development 2–4; potential benefits of
33; power in 224–9; regulation in
34–47, 224–9; as responsible,
sustainable, ethical 1–2, 34–47, 46–7,
224–9; as route out of poverty 53; self-
regulation of 34–7; and student
'gappers' 227–8; unequal/uneven
nature of 60
Tourism Concern 32, 35, 41, 47, 65, 66,
68, 70, 74, 158
tourism development, air/sea boom 13;
changing nature of 10–11; colonial
legacy 11–12; concept 2, 16–17; and
conquistadors 11; and dependency
theory 12; and environmental/
ecological concerns 13–14; and
explorers/backpackers 12–13; and First
World/Third World relationship 13–15;
and focus on poverty 58; importance of
21; local participation on 75–7; and
military control 12; and Millennium
Development Goals 17–21; and
neoliberal economics 14–15; and
physical/economic infrastructure 21;
and plantation/slave system 11;

political nature of 3, 11–16; and
poverty/inequality 17; and solidarity
tours 13; sustainable 3; western
capitalist view 16–17
Tourism Link journal 79
Tourism and Travel-Related Services 31
transnational corporations (TNCs) 10–11,
21, 25, 27, 29, 37
trickle-down effects, case studies 83–94; in
cities 177, 187
Tropic Ecological Adventures (Ecuador)
61, 82
Truman, Harry S. 2, 16
TUI Group 218
TUI Nederland 38
TUI Nordic 208, 218

UK Emissions Trading Scheme 121
UN-HABITAT 172
UNESCO 38, 79, 106, 180
United Nations Development Programme
(UNDP) 19, 56
United Nations Environment Programme
(UNEP) 38, 108, 113, 115, 226
United Nations General Assembly 40
United Nations High Commissioner for
Human Rights (UNHCR) 56
United Nations Human Rights Committee
92
United Nations Millennium Summit
(2000) 17
United Nations Population Fund (UNFPA)
210, 220
United Nations (UN) 14, 41, 54, 60, 172,
226
United Nations Working Group on
Contemporary Forms of Slavery 204
United States Agency for International
Development (USAID) 164
Urban and Local Government Strategy
(World Bank, 2000) 178–9
urban tourism 7; centrality of 172; and
cultural heritage 180–2; development
of 175–6; and *favela* tours 186–8;
impact of 183–4, 189–91; and pro-poor
tourism 182–6
US National Parks System 122

Vallejo, J.R. 131
Venta Club 36
Vidal, J. 14
Villa Boas brothers 141–2
Viner, D. 118

Wadhams, N. 217
Washington Consensus 21, 22, 46, 57, 177
WaterAid 174
websites 228–9, 231–3
Wells, M. 77
Welsh, P. 196
West, P. 77
Wheeller, B. 36
Williams, M. 28
Wolfensohn, James 20
Wooding, B. 22, 58
Woolcock, 174
World Bank 7, 10, 17, 21, 24, 32, 53, 54,
    57, 76, 92, 115, 122, 141, 178–9, 185
World Commission on Environment and
    Development 108
World Conservation Strategy 108
World Development Movement (WDM)
    10, 29, 31, 32
World Development Reports 17, 54
World Health Organisation (WHO) 212
World Heritage Committee 106
World Heritage List 180–1, 182

World Parliamentary Summit 20
World Social Forum 225, 229
World Tourism Organisation (WTO/OMT)
    7, 14, 19, 21, 28, 32, 37, 38, 40, 41, 53,
    109–10, 119, 216–17, 225
World Trade Organisation (WTO/OMC)
    14, 19, 22, 28, 29, 31, 32, 41, 45, 53,
    57, 92, 225
World Travel and Tourism Council
    (WTTC) 36, 37, 40, 53
World Travel and Tourism Environment
    Research Centre (WTTERC) 37
World Urban Forum 175
World Vision 206
World Wide Fund (WWF) for Nature 35,
    47, 76, 93, 108, 113, 117, 122

Xcaret (Mexico) 129
Xel-há (Mexico) 129
Xingu Indian Land Association 142
Xingu National Park (Brazil) 141–2

Zambrana, E. 214